DIFFERENT SECTORS OF MACHINE LEARNING AND AI

DIFFERENT SECTORS OF MACHINE LEARNING AND AI

Dr. Sanjay B. Waykar
Dr. Anil W. Kale
Dr. Nandkishor P. Karlekar
Dr. Vaibhav E. Narawade

NAMYA PRESS.com

Published by

Namya Press

NAMYA PRESS.com

USA: 10685-B Hazlehurst Dr. # 31013 Houston, Texas - 77043
Email: namyapress@gmail.com
Website: https://namyapress.com

Edition	:	First Published in 2022 (Namya Press)
Title	:	Different Sectors of Machine Learning and AI
Editors	:	Dr. Sanjay B.Waykar, Dr. Anil W. Kale, Dr. Nandkishor P. Karlekar & Dr. Vaibhav E. Narawade
ISBN	:	978-1-956861-23-5

Preface

It's with great happiness that I would like to acknowledge a great deal of people that get helped me extremely through the entire difficult, challenging, but a rewarding and interesting path towards some sort of Edited Book Without having their help and support, none of this work could have been possible.

Dr. Sanjay B. Waykar, Dr. Anil W. Kale, Dr. Nandkishor P Karlekar, Dr. Vaibhav Narawade, we all forever indebted to my father and mother and also the rest of my children. My wife is a huge great source of inspiration in my experience. None of this will have been achieved without her really like, support and steady encouragement and Also Support My All National and International Friends.

My parents' prayers have always accompanied me. Their love keeps me ambulatory. I am grateful to all of them.

Contents

Chapter - 01

Intelligent Covid-19 Detection Using Machine Learning

Dr. Sanjay B. Waykar

Mahatma Gandhi Mission's College of Engineering and Technology, India.
Email:sbwaykar@gmail.com

Dr. Anil W. Kale

Mahatma Gandhi Mission's College of Engineering and Technology, India.
Email: anil5474@gmail.com

Dr. Nandkishor P. Karlekar

MGMCET Kamothe Navimumbai, India. Email: npkarlekar.mgmcet@gmail.com

Dr. Vaibhav Narawade

Ramrao Adik Institute of Technology, India. Email:vnarawade@gmail.com

INTRIDUCTION

In December 2019, an original serious infectious respiratory condition Covid 2, which is a sort of Severe Acute Respiratory Syndrome (SARS-CoV-2) infection called COVID-19, was found in Wuhan, China. Coronavirus infection is airborne and can without much of a stretch spread and contaminate individuals. As per the Centers for Disease Control and Prevention (CDC), the tainted individuals show a scope of indications like dry hack, windedness, weariness, losing the feeling of taste and smell, the runs, and clog Tainted patients can likewise introduce fever scenes. Surprisingly, a few patients who have gotten the infection probably won't show any of the previously mentioned side effects. They can feel totally ordinary conveying the infection and proceeding to spread the illness without knowing. As COVID-19 has a quick nature of spreading, the World Health Organization (WHO) proclaimed it as a worldwide pandemic in March 2020. At the hour of composing this invention (i.e., September 2020), the absolute number of affirmed COVID-19 cases overall was more than 32 million.

To handle this episode, researchers in various examination networks are looking for a wide assortment of PC helped frameworks like the Internet of Things Machine Learning (ML) or Deep Learning (DL) procedures. These innovations can be utilized for controlling the spread of the infection, distinguishing the infection, or in any event, planning and assembling an immunization or medication to battle it. There were two scourges in the past from the Covid family including Severe Acute Respiratory Syndrome (SARSCoV) and Middle Eastern Respiratory Syndrome (MERS).

SARS-CoV is a respiratory infection that was contagious from one individual to another and it was first distinguished in 2003. The infection had more than 8,000 affirmed cases overall during its course which influenced more than 26 nations MERS is likewise a respiratory infection with comparable side effects of SARS-CoV. ML, as a subset of Artificial Intelligence (AI), has shown a great deal of possibilities in numerous ventures. ML procedures can be customized to emulate human insight. For instance, in the medical services industry, ML procedures can be prepared and utilized towards clinical finding. ML models have been unfathomably prepared over a dataset comprising of clinical pictures like Computed Tomography (CT) Scan, Magnetic Resonance Imaging (MRI), peculiarities.

During past pandemics, ML procedures have been broadly carried out to help medical care experts for better activities in regards to the illnesses. For instance, a ML model that uses GPS innovation alongside distributed computing force and Google Maps to address conceivably contaminated patients and give an elective course to uninfected clients coming about in possibly moderating the spread. The model arrives at the characterization exactness of 80% in re-directing away from tainted patients. In another investigation, Choi et al. [30] utilized ML models for nostalgic investigation to survey public overcompensation showing up in media articles and web-based media stages.

This sort of inside and out investigation can quickly screen the public response. It can likewise help policymakers in making the right moves in diminishing trepidation and pain from general society in regards to MERS. ML has additionally been broadly utilized to further develop clinical dynamic in regards to the current COVID-19 pandemic. Specialists, utilizing ML calculations and grouping strategy, can figure the spread in regions. ML techniques for picture characterization are utilized by established researchers to help in diagnosing the dangerous infection. With the goal of discovering a remedy for the infection, ML calculations are utilized to assess how reliable are off-the-counter medications might be utilized to help tainted patients.

The first Covids found on the planet were liable for respiratory and intestinal contaminations, of which by far most had a self-restricted course and drove generally to manifestations of normal virus. Be that as it may, they can ultimately form into genuine contaminations in bunches in danger (heart infections, diabetes, among others), in the old and furthermore in kids. Prior to the start of this current pandemic, two profoundly pathogenic Covid species (SARS and MERS) were portrayed and were liable for flare-ups of extreme intense respiratory conditions.

Concerning new (COVID-19) it was perceived as a causative specialist of pneumonia that prompts extreme intense respiratory disorder (SARS-CoV-2). One of its principle challenges is its quick bandwidth and, at times, movement to serious pneumonic conditions that have requested from the wellbeing framework a consideration and battle technique never found in the entire world. In Brazil, the assumption is to be of dramatic development, which is the reason there is a need to execute extreme measures to control populace course and counteraction. Because of the speed of transmission, in many nations, including Brazil, beginning phase preventive measures were not carried out, causing a blast of indicative cases, a large number of them serious, with delayed interest from tertiary wellbeing administrations.

Given this situation, the inescapable rise of an enormous unexpected of basically sick patients with COVID-19, with various guesses, made it essential to look for early demonstrative systems for better screening and therapy sufficiency for each situation. In this unique circumstance, even before affirmation of the contamination, evaluating for patients with respiratory indications is brought out through clinical examination and imaging tests like Chest Computed Tomography (CT).

In clinical assessment, the principle depicted manifestations of contamination are fever (88.5%), hack (68.6%), myalgia or weakness (35.8%), sputum (28, 2%) and dyspnea (21.9%). Different indications likewise depicted incorporate cerebral pain and tipsiness (12.1%), loose bowels (4.8%), queasiness and spewing (3.9%). Furthermore, some haematological changes were noticed: lymphocytopenia (64.5%), expanded C-responsive protein (CRP) (44.3%), expanded lactic dehydrogenase (DHL) (28.3%), and leukopenia (29, 4%). Chest CT is thinking about as the best imaging technique for appraisal of COVID-19, since customary radiography has low affectability, outstandingly in beginning phases. Regular discoveries portrayed in the writing incorporate ground-glass opacities (GGO) with a more fringe appropriation, related with septal thickening and solidifications, generally influencing various projections, albeit these discoveries can likewise be found in other viral pneumonias.

Along these lines, AI applied to symptomatic imaging can empower the improvement of apparatuses that can normalize the determination and give potential discoveries reminiscent of the presence of the infection, its seriousness, and accordingly its guess. Since the start of the pandemic, because of its crisis, a few examinations opened up to attempt to foresee most noticeably terrible results. Fundamental conceivable danger factors were assessed in review examines. A large portion of these investigations show old age, weight and other comorbidities (diabetes, serious asthma and other respiratory sicknesses, heart, liver, neurological and kidney infections and immune system illnesses) as the principle players for a most noticeably terrible result.

AI has been utilized by a developing number of studies in this situation and in other wellbeing related fields, going from assisting with analysis until giving more vigorous proof to asset distribution upheld that further examinations are need to show all the capability of this apparatus in clinical practice. The significant benefit of utilizing AI is that we can join various factors (segment and clinical information, lab examines and imaging) in a huge scope, with diminished paces of misdiagnosis and having the option to give productive bits of knowledge in a few parts of the sickness. In our convention, we will extrapolate the current craftsmanship by

1. The acquiring information from many patients hospitalized in nine distinctive private and public foundations in Brazil
2. The remove the human blunder and the high between rater concession to the assessment of chest CTs,
3. The predict the likelihood of various results (Time to clinic release; length of stay in the ICU; orotracheal intubation because of intense respiratory disappointment; improvement of intense respiratory uneasiness disorder, and furthermore other optional results that will be portrayed later in this convention).

OBJECTIVES

- The objective of the invention is to provide a assess potential changes in Chest CT, through a score, that propose a more terrible anticipation in patients with COVID-19, and to distinguish designs corresponded with more awful clinical turns of events, to direct, in the forthcoming unfurling of the investigation, the assessment of prognostic markers emerging computerized examinations of Chest CT and add to focusing on treatment as indicated by seriousness (orotracheal intubation, hospitalization).

- The other objective of the invention is to provide a information base with clinical pictures and their individual anonymized reports for CT methodology, in various transform utilitarian changes, in patients with intense respiratory disorders.
- The other objective of the invention is to provide a Evaluate the exhibition of AI calculations in this information for undertakings like grouping, division, picture enrollment and understanding of reports.
- The other objective of the invention is to provide a Evaluate the effect of the utilization of these models on clinical act of imaging experts.

MATERIALS AND METHODS

We will direct a reflectively longitudinal multicentre study (9 Institutions) with something like 160 patients hospitalized from March to May 2020 because of clinical signs and manifestations of intense respiratory condition. This investigation was supported by our National Ethics committeeand endorsed across each taking part place's morals advisory group (Universidade Federal do Rio de Janeiro, Universidade do Estado do Rio de Janeiro, Universidade Federal de São Paulo, Hospital 9 de julho, Hospital São Lucas, Hospital Santa Paula, Hospital Alemão Oswaldo Cruz). CONEP is the focal morals panel. We mentioned waiver of the agree structure because of the review study plan.

Study Populace

Qualified patients for the examination should meet the accompanying attributes will be considered qualified for the investigation:

1. Signs and side effects of intense respiratory s
2. Syndrome
3. Positive epidemiological history for COVID-19, which might incorporate ongoing contact (most recent 14 days) with an affirmed or suspected case, late excursion (most recent 14

days) to a high-occurrence area, or show of indications after the beginning of the local area transmission period of SARS-CoV-2 (after 3/20/2020) when the date of hospitalization.

4. Have performed, when suggestive, a chest processed tomography.

Clinical Imaging

Diagnosing COVID-19 is quite possibly the main pieces of managing the sickness. Because of low access and high chance of bogus adverse outcomes to the RT-PCR packs, there is a fundamental requirement for utilizing different methodologies, for example, clinical pictures examination for exact and dependable screening and conclusion in COVID-19. As a general rule, dissecting clinical imaging modalities, for example, chest X-beam and CT-Scan have key commitments in affirming the conclusion of COVID-19 just as screening the movement of the illness. Distinctive ML methods that fuse X-beam and CT-Scan picture preparing approaches could help doctors and medical services experts as a superior way for finding and comprehension of the movement of the COVID-19 illness.

1) X-beam: During this pandemic, chest imaging can be a significant piece of the COVID-19 in beginning phase of recognition. Arranging patients quickly is what is generally anticipated from these methodologies. Inside the order of clinical imaging, Chest X-Ray (CXR) was prescribed to be executed as the primary clinical imaging in regards to COVID-19 by the Italian Society of Radiology (SIRM.). CXR has an affectability of 67.1% which can be first executed in quite a while incorporating helping radiologists with better COVID-19 cases recognizable proof and quick treatment allocating to the patient. Moreover, CXR is cheap and secure in light of limiting the danger of defilement which makes a more secure working environment for medical services labourers also.

To diminish the measure of work by radiologists, ML procedures can be relegated to arrange patients concerning COVID-19. To do that, specialists are generally centered around the ML arrangement models like Support Vector Machine (SVM), Convolutional Neural Networks (CNN), DL. One methodology executed X-Ray to arrange the lung injuries (brought about by COVID-19) with Multi-Level Threshold (MT) cycle and SVM model.

Inside this model, initially, the lung picture difference will be improved. Besides, the picture will be diminished into explicit segments (utilizing MT) to keep away from duplication of work on uninfected regions. Ultimately, the SVM model characterizes the areas of the lung concerning the predefined solid lungs. The fostered a stage utilizing an assortment of Deep Convolutional Neural Networks (DCNN) models characterizing inside the SVM with two unique datasets to recognize COVID-19 cases dependent on the connected CRX picture. The proposed a DCNN model utilizing the information assembled from two emergency clinics in Italy to addresses the significance of AI in the identification of COVID-19. The prepared ML models over an enormous viral pneumonia dataset of CXR pictures to identify inconsistencies. They tried their model on a totally unique dataset that has COVID-19 CXR pictures.

This is done as one of the side effects of COVID-19 can be pneumonia. The outcomes are amazing as the model performs well when tried on the COVID-19 dataset with the Area Under the Curve (AUC) of 83.61%. It is considerably more amazing as the model was prepared on an alternate dataset but performed well. Likewise, used COVIDx dataset, an openly accessible dataset comprises of COVID-19, pneumonia and non-COVID-19 pneumonia related X-beam pictures. The creators utilized this information to prepare their model for recognition of COVID-19, the Deep Neural Network (DNN) is alluded to as COVID-Net appearance encouraging outcomes in diagnosing contaminated patients. utilized exchange learning approaches like element extraction and calibrating of CNN based models and prepared and tried over comparable datasets accomplishing a forecast exactness up to practically 98%.

They showed that executing move learning can have a huge improvement in outcomes. Most ML classifiers are prepared and tried to accomplish high expectation exactness of COVID-19; in any case, evaluate the vulnerability that could exist by utilizing such classifiers as an essential mechanism of conclusion. A way to deal with approve the ML expectation of conclusion in CXR pictures was evaluated. It took advantage of a Bayesian Deep Learning classifier to assess the model vulnerability. The outcome examination shows a solid relationship amongst vulnerability and precision of expectation, which implies that the higher the vulnerability result, the more dependable the forecast exactness.

Chatbots

PC programs created to speak with people by embracing normal dialects are called chatbots. A chatbot fundamentally can speak with various clients and produce appropriate reactions to those client's dependent on their bits of feedbacks. As of late, the COVID-19 pandemic has prompted constructing diverse chatbots as opposed to utilizing hotlines as a specialized technique. This will decrease medical clinic visits and increment the proficiency of conveying. By and large, chatbots are executed to give an online discussion the client by one or the other content or voice shows on web applications, cell phone applications, channels, and This discussion can assist the client with having a superior comprehension of their circumstance and gives a few clues to clients so the person can make legitimate strides. Chatbots are generally considered as extraordinary compared to other fit to screen patients distantly without com.

The benefits of them incorporate rapidly refreshing data, tediously uplifting new practices like washing hands, and helping with mental help because of the pressure brought about by separation and falsehood. The ML-based chatbots are improved during the preparation methodology and utilizing more information makes this methodology more solid. During the COVID-19 pandemic, chatbots are standing out enough to be noticed to give more insights concerning COVID-19 in various stages. A wide assortment of chatbots with various dialects have been carried out to help patients at the beginning phase of COVID-19. “Aapka Chkitsak”, an AI-based chatbot created by in India, helps patients with far off meeting in regards to their wellbeing data, and medicines.

The new Covid, which started to be called SARS-CoV-2, is a solitary abandoned RNA beta Covid, at first recognized in Wuhan (Hubei area, China) and right now spreading across six landmasses making a significant mischief patient, with no particular instruments as of recently to give prognostic results. Consequently, the point of this investigation is to assess potential discoveries on chest CT of patients with signs and manifestations of respiratory disorders and positive epidemiological variables for COVID-19 contamination and to associate them with the course of the sickness. In this sense, it is additionally expected to foster explicit AI calculation for this reason, through aspiratory division, which can foresee conceivable prognostic elements, through more exact outcomes.

Our elective theory is that the AI model dependent on clinical, radiological and epidemiological information will actually want to foresee the seriousness forecast of patients contaminated with COVID-19. We will play out a multicenter review longitudinal examination to acquire countless cases in a brief timeframe, for better investigation approval. Our comfort test (something like 20 cases for every result) will be gathered in each middle thinking about the consideration and rejection standards. We will assess patients who enter the medical clinic with clinical signs and side effects of intense respiratory condition, from March to May 2020.

We will incorporate people with signs and manifestations of intense respiratory condition, with positive epidemiological history for COVID-19, who have played out a chest processed tomography. We will evaluate chest CT of these patients and to connect them with the course of the sickness. Essential results:

1. Time to emergency clinic release;
2. Length of stay in the ICU;
3. Orotracheal intubation;
4. Development of Acute Respiratory Discomfort Syndrome. Auxiliary results:
    - Sepsis;
    - Hypotension or cardiocirculatory brokenness requiring the solution of vasopressors or inotropes;
    - Coagulopathy;
    - Acute Myocardial Infarction;
    - Acute Renal Insufficiency;
    - Death. We will utilize the AUC and F1-score of these calculations as the fundamental measurements, and we desire to recognize calculations fit for summing up their outcomes for each predetermined essential and optional result.

DIAGRAM

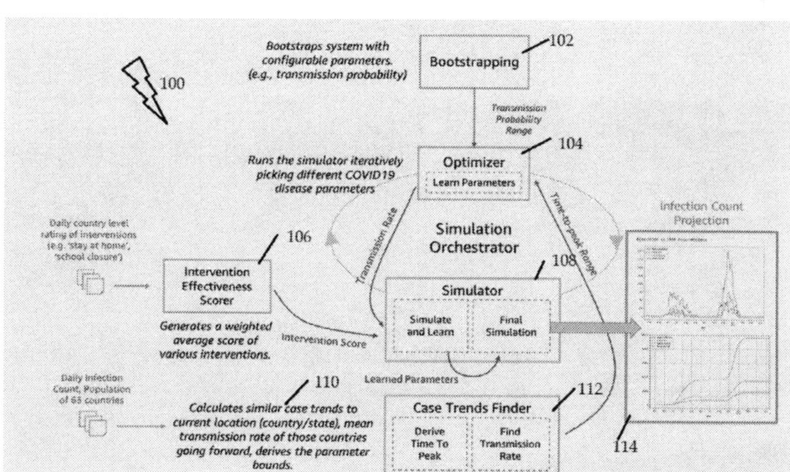

Fig.1: COVID-19 Detection, Spread Prevention and Medical Assistance Using Machine Learning Flow Chart.

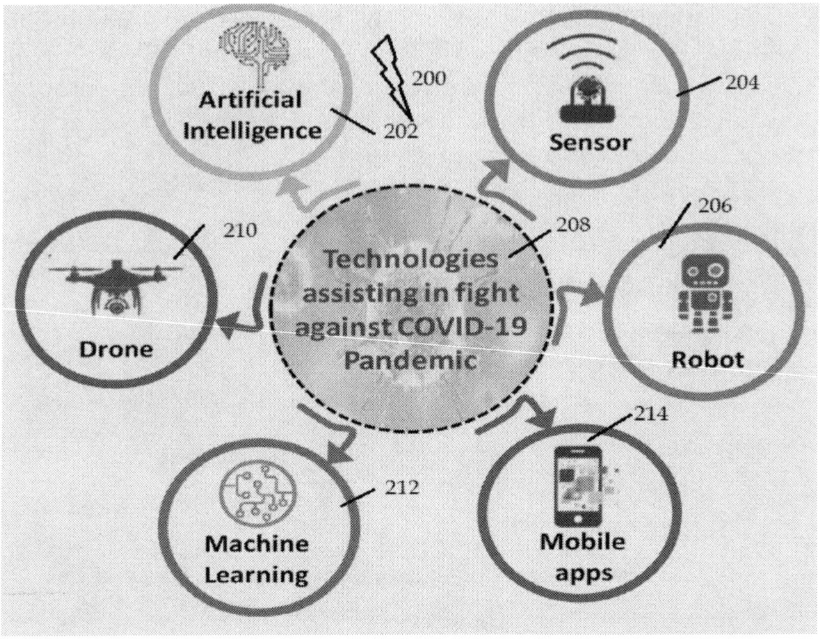

Fig. 2: COVID-19 Detection, Spread Prevention and Medical Assistance Using Machine Learning Block Diagram.

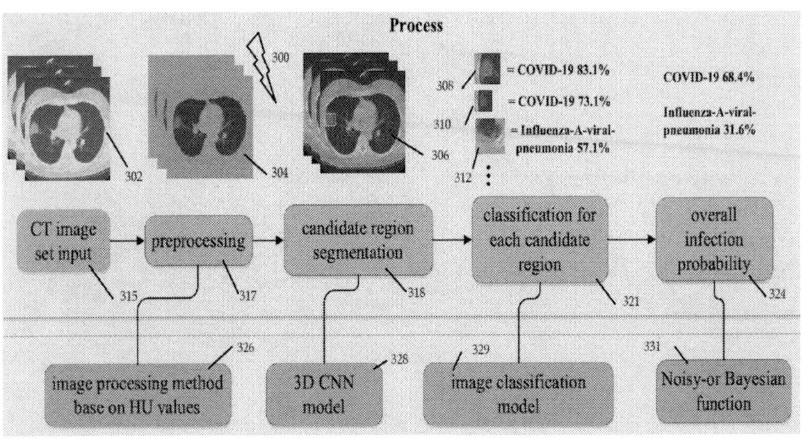

Fig. 3: COVID-19 Detection, Spread Prevention and Medical Assistance Using Machine Learning

METHODS

Our Invention COVID-19 Detection, Spread Prevention and Medical Assistance Using Machine Learning is a Coronavirus was first found in December 2019 and has proceeded to quickly spread across nations overall tainting thousands and millions of individuals. The infection is lethal, and individuals who are experiencing earlier sicknesses or are more established than the age of 60 are at a higher danger of mortality. Medication and Healthcare businesses have flooded towards discovering a fix, and various approaches have been altered to alleviate the spread of the infection. While Machine Learning (ML) techniques have been broadly utilized in different areas, there is presently a popularity for ML-helped conclusion frameworks for screening, following, and anticipating the spread of COVID-19 and discovering a fix against it. In this innovation, we present an excursion of which job ML has played so far in fighting the infection, basically taking a gander at it from a screening, estimating, and immunization points of view. We present an extensive overview of the ML calculations and models that can be utilized on this undertaking and help with fighting the infection. Medication and Healthcare businesses have flooded towards discovering a fix, and various strategies have been corrected to relieve the spread of the infection. While Machine Learning (ML) techniques have been generally utilized in different spaces, there is presently a popularity for ML-helped analysis frameworks for screening, following, and anticipating the spread of COVID-19 and discovering a fix against it. In this invention, we present an excursion of which job ML has played so far in fighting the infection, fundamentally taking a gander at it from a screening, gauging, and antibody points of view. We present an exhaustive review of the ML calculations and models that can be utilized on this campaign and help with engaging the infection. Record Terms—COVID-19, Machine Learning, Artificial Intelligence, Healthcare, Drug Development, Prevention, Predictive Analysis, Diagnosis, Image Classification.

DESCRIPTION

Artificial Intelligence of Things (AIoT)

As a rule, uses of the Internet of Things (IoT) and AI can help organizations with measure mechanization which would diminish the contacts of people because of the lower number of individuals required. During the COVID-19 pandemic, AI and IoT are standing out enough to be noticed in the medical care space where screening and distinguishing systems should be possible all the more securely. Warm imaging and social distance observing are two primary capacities that are mostly thought to be in the screening period of COVID-19. Truth be told, the points of those gadgets are high-temperature discovery, face veil screening, and distance controlling that are talked about in the coming areas.

Warm Imaging

Concerning the IoT warm screening applications, AI can aid this region by carrying out proper calculations. SmartX, a warm screening gadget utilizing infrared warm imaging and AI face acknowledgment makes separating swarm structures or doorways more effective (see Figure 3). The gadget catches a guest's temperature and furthermore checks whether the person in question is wearing a face veil.

Social Distance Monitoring

Regarding the need of rehearsing social removing utilizing IoT gadgets, AI can carry out a mechanized screening utilizing PC vision techniques. An occurrence of such a gadget is RayVision which guarantees social separating and face veil wearing rules are continued in the group. By utilizing the PC vision methods, it can screen individuals with a live stream on its particular dashboard which permits cautioning the experts if there should arise an occurrence of any standard breaking.

The Similarly, Landing AI is another AI-based innovation which can recognize social removing infringement continuously. Also, a friend investigated research executed an Unmanned Aerial Vehicle (UAV) or a robot with the ML application in light of the requirement for keeping social separation in swarms. Curiously, veils will be given. Signs, Preventing the Spread Obtaining early-cautioning finishes invention work for a flare-up of a plague could truly help towards easing back and moderating the spread of the infection. It additionally can urge social orders to take important prudent steps. In this segment, we audit the early-cautioning signs that were made conceivable utilizing the ML innovation.

The World Health Organization (WHO) offered expressions about COVID-19 being an expected episode on the ninth of January 2020. There were AI organizations like BlueDot and Metabiota that had the option to foresee the episode considerably prior. BlueDot centers around spotting and anticipating flare-ups of irresistible sicknesses utilizing its exclusive strategies and apparatuses. They use ML and NLP strategies to channel and center the danger of spreading an infection. Utilizing the information from nearby news reports of initial not many associated cases with COVID-19, authentic information on creature infection flare-ups, and carrier ticket data, they were then ready to utilize their apparatus to anticipate an unmistakable flare-up happening inside approaching urban communities and different districts of China.

BlueDot had cautioned its customers about the episode on the 31st December 2019, longer than seven days before the WHO offered any expressions about it. Also, Metabiota utilized their ML calculations and Big Data to anticipate flare-ups and spreads of illnesses, and occasion seriousness. They utilized their innovation and flight information to foresee that there will be a COVID-19 flare-up in nations like Japan, Thailand, Taiwan, and South Korea.

Contact Tracing

The significant methodologies for forestalling the spread of the infection is following the affirmed instances of COVID-19 due to the expected spread of the infection through beads by hacks, wheeze, or talking. It is suggested that not just individuals who have a positive COVID-19 test, yet additionally the ones who had been in close contact with the affirmed cases to be isolated for 14 days. contact following applications are applied all around the world for this reason with various techniques. Fundamentally, it begins after the analysis interaction on the grounds that the identified case should be followed. In particular, after the information is gathered by those applications, ML and AI procedures will begin examining for finding additionally spread of the sickness.

Albeit the contact following applications could be profoundly useful during the pandemic, protection issues can bring high concerns with respect to the reconnaissance of people by certain legislatures because of immense measure of the gathered information. Utilizing the computerized impression information furnished by the applications alongside ML innovation could permit clients to recognize contaminated patients and implement social removing measures.

Estimating

Estimating scourges fixates on following and foreseeing the spread of irresistible sicknesses and infections. During a pandemic, anticipating techniques and models can be prepared on epidemiological related information to give an expected number of contaminated cases, examples of spread that can give medical care labourers direction on the best way to get ready suitably for an episode. Beforehand measurable gauging device, for example, Susceptible, Infected, recovered models (SIR Models) have been utilized to decide the spread of a sickness through the populace Recently, with the COVID-19 pandemic.

Online Media Analysis Social

Media has become a stage where individuals share pictures, surveys, posts, and trade stories. A famous online media stage where individuals might get and get to news is Twitter. Its clients can approve live cautions and acquire data straightforwardly through the cell phone application. This is conceivable as significant media sources, government bodies, public venues, and so on, all have accounts that they use to share reports on Twitter. Clients can likewise utilize the stage to share their own encounters by means of tweets.

Understanding the Virus

Breaking down the genomics and proteomics attributes of a viral sickness is a significant advance to battle the infection. Researchers have been contemplating the virology of COVID-19 which can give the physical and substance properties, cell section and receptor connection, and the general biology and the genomics of the infection. A genome is the finished hereditary data that gives the design of an infection and knowing the genome for COVID-19 can help in better understanding the contagiousness and irresistibleness of the infection [186]. The investigation of proteomics is knowing the proteins of an organic entity.

Recognizing the proteins of COVID-19 would permit a superior comprehension of the general protein structure and find how the proteins would communicate with the medications. Over late years, there have been noteworthy progressions by researchers in interdisciplinary fields of bioinformatics and computational medication and ML methods have shown significant translation towards deciding genomics and protein designs of different infections. In this segment, we center around COVID-19 and talk about the ML methods that have been carried out with respect to the exploration of deciphering the genomics and proteomics of that.

Medication and Vaccine Development

As COVID-19 cases keep on rising quantities of both the tainted cases and the loss of life, it has become an earnest need to find a medication that could moderate these numbers from expanding any further. ML procedures can be utilized to dissect how medications respond to the viral proteins of COVID-19. We have effectively seen ML strategies and procedures like SVM and Bayesian Classifiers being utilized for drug disclosure and repurposing [198]. In this segment we survey the ML studies and exploration that had been done about finding the new medications or repurposing the right now supported FDA ones. We likewise audit the ML research that has been finished in regards to the antibody advancement.

Drug Development and Repurposing

An exploratory methodology of deciding if monetarily accessible antiviral medications can treat or help towards diminishing the seriousness of COVID-19 tainted patients was introduced. They utilized a pre-prepared ML learning model called Molecule Transformer-Drug Target Interaction (MT-DTI), a connection expectation model, to anticipate the limiting liking between COVID-19 tainted proteins and mixtures. The goal of their examination was to recognize potential FDA endorsed drugs that might control the proteins of COVID-19. MT-DTI is equipped for foreseeing the substance successions and amino corrosive arrangements of an objective protein without the entire construction data. This is useful to use as there was restricted information on the general construction of viral proteins of COVID-19 at first. Concerning advantage, the creators utilized the MT-DTI model to foresee restricting affinities of 3,410 FD-Approved medications.

Also, utilized exclusive DL methods with the end goal of medication disclosure. They utilized their model to assess how FDA and European Medicines Agency (EMA) endorsed medications and mixtures would influence human cells, dissecting more than 1,660 medications.

Immunization Development

When an infection begins to spread and transform into a worldwide pandemic, there is a tiny shot at halting it without an immunization. That stands valid for COVID-19 also. Verifiably, inoculation has been the answer for control or slow the spread of a viral disease. It is basic to have an antibody created to give resistance against COVID-19 and stop this pandemic. Up until now, the exploration for antibody advancement of COVID-19 is committed with three unique kinds of vaccines. The Whole Virus Vaccine addresses a traditional system for the improvement of inoculations of viral sickness. Subunit Vaccine depends on separating the insusceptible reaction against the S-spike protein for COVID-19. This will abstain it from docking it with the hosts receptor protein. The Nucleic Acid Vaccine delivers a defensive immunological reaction to battle against the infection by.

OUTCOME

While Machine Learning (ML) techniques have been broadly utilized in different areas, there is presently a popularity for ML-helped conclusion frameworks for screening, following, and anticipating the spread of COVID-19 and discovering a fix against it. In this innovation, we present an excursion of which job ML has played so far in fighting the infection, basically taking a gander at it from a screening, estimating, and immunization points of view. We present an extensive overview of the ML calculations and models that can be utilized on this undertaking and help with fighting the infection. Medication and Healthcare businesses have flooded towards discovering a fix, and various strategies have been corrected to relieve the spread of the infection. While Machine Learning (ML) techniques have been generally utilized in different spaces, there is presently a popularity for ML-helped analysis frameworks for screening, following, and anticipating the spread of COVID-19 and discovering a fix against it. In this invention, we present an excursion of which job ML has played so far in fighting the infection, fundamentally taking a gander at it from a screening, gauging, and antibody points of view. We present an exhaustive review of the ML calculations and models that can be utilized on this campaign and help with engaging the infection. Record Terms—COVID-19, Machine Learning, Artificial Intelligence, Healthcare, Drug Development, Prevention, Predictive Analysis, Diagnosis, Image Classification.

- According to claim1# the invention is to a While Machine Learning (ML) techniques have been broadly utilized in different areas, there is presently a popularity for ML-helped conclusion frameworks for screening, following, and anticipating the spread of COVID-19 and discovering a fix against it. In this innovation, we present an excursion of which job ML has played so far in fighting the infection, basically taking a gander at it from a screening, estimating, and immunization points of view.
- According to claim1,2 # the invention is to a present an extensive overview of the ML calculations and models that can be utilized on this undertaking and help with fighting the infection. Medication and Healthcare businesses have flooded towards discovering a fix, and various strategies have been corrected to relieve the spread of the infection. While Machine Learning (ML) techniques have been generally utilized in different spaces, there is presently a popularity for ML-helped analysis frameworks for screening, following, and anticipating the spread of COVID-19 and discovering a fix against it.

- According to claim1,2,3# the invention is to a we present an excursion of which job ML has played so far in fighting the infection, fundamentally taking a gander at it from a screening, gauging, and antibody points of view. We present an exhaustive review of the ML calculations and models that can be utilized on this campaign and help with engaging the infection. Record Terms—COVID-19, Machine Learning, Artificial Intelligence, Healthcare, Drug Development, Prevention, Predictive Analysis, Diagnosis, Image Classification.

REFERENCE

[1] Abhigna, B.S., Chakraborty, S., Kumar, V., Medha: A collaborative E-concept framework, Proceedings of the 2018 International Conference On Communication, Computing and Internet of Things, IC3IoT 2018, 2019, pp. 117–122, 8668127

[2] Abidin, S., Vadi, V.K., Rana, A., On Confidentiality, Integrity, Authenticity, and Freshness (CIAF) in WSN, Advances in Computer, Communication and Computational Sciences, 2020, pp 87-97

[3] Ahay, A., Choudhury, T., Kumar, V., Hierarchical Evolution of Digital Games, Proceedings of 2019 International Conference on Computational Intelligence and Knowledge Economy, ICCIKE 2019, 2019, pp. 381–386, 9004378

[4] Area: Communication, Frequency modulation, Networks, wireless communication

[5] Area: Communication, Frequency modulation, Networks, wireless communication

[6] Area: Communication, Frequency modulation, Networks, wireless communication

[7] Area: Communication, Frequency modulation, Networks, wireless communication

[8] Arora, M., Kumar, V., Dave, M., Hybrid cloud infrastructure for load balancing and scaling with vm path optimization using ant colony optimization, Journal of Advanced Research in Dynamical and Control Systems, 2020, 12(5 Special Issue), pp. 660–674

[9] Arora, M., Kumar, V., Dave, M., Task scheduling in cloud infrastructure using optimization technique genetic algorithm, Proceedings of the World Conference on Smart Trends in Systems, Security and Sustainability, WS4 2020, 2020, pp. 788–793, 9210303

[10] Bansal, R., Gupta, A., Singh, R., Nassa, V.K., Role and impact of digital technologies in E-learning amidst COVID-19 pandemic, Proceedings - 2021 4th International Conference on Computational Intelligence and Communication Technologies, CCICT 2021, 2021, pp. 194–202

[11] Bansal, R., Obaid, A.J., Gupta, A., Singh, R., Pramanik, S., Impact of Big Data on Digital Transformation in 5G Era, Journal of Physics: Conference Seriesthis link is disabled, 2021, 1963(1), 012170

[12] Chauhan, S.S., Khare, A.K., Kinematic analysis of the ABB IRB 1520 industrial robot using roboanalyzer software, Evergreenthis link is disabled, 2020, 7(4), pp. 510–518

[13] Chauhan, S.S., Khare, A.K., Method of Obtaining Planar State of Stress Using Mohr's Circle—Some Typical Cases, Lecture Notes in Mechanical Engineering, 2022, pp. 33–38

[14] Goel, S., Tripathi, A., Choudhury, T., Kumar, V. Parkinson's Disease Detection using Soft Computing Technique, Proceedings of the 2019 8th International Conference on System Modeling and Advancement in Research Trends, SMART 2019, 2020, pp. 3–6, 9117384

[15] Govindarajan, V., Sivakumar, R., Patil, P.P., Kannan, M., Ramesh, B., Effect of Tungsten Carbide Addition on the Microstructure and Mechanical Behavior of Titanium Matrix Developed by Powder Metallurgy Route, Advances in Materials Science and Engineeringthis link is disabled, 2022, 2022, 2266951

[16] Gupta, A., Singh, R., Nassa, V.K., Sharma, P., Koti, K., Investigating Application and Challenges of Big Data Analytics with Clustering, 2021 International Conference on Advancements in Electrical, Electronics, Communication, Computing and Automation, ICAECA 2021, 2021

[17] https://journals.plos.org › plosone › article › journal.po

[18] https://pubs.rsna.org › doi › pdf › radiol.2020200905

[19] https://www.drdo.gov.in › ai-based-intelligent-covid-19

[20] https://www.nature.com › nature communications › articles

[21] https://www.ncbi.nlm.nih.gov › articles › PMC8071762

[22] Jain, P., Vijay, S., & Gupta, S. C. (2009). ATM Network QoS Parameters and Performance Analysis using OPNET Simulator. International Journal of Recent Trends in Engineering, 2(4), 228.

[23] Jain, P., Vijay, S., & Gupta, S. C. Performance Analysis & QoS Guarantee in ATM Networks. Global Journal of computer Science and Technology, 131-136.

[24] Jain, S., Choudhury, T., Kumar, V., Kumar, P., Detecting malware and analysing using sandbox evasion, Proceedings of the 2018 International Conference On Communication, Computing and Internet of Things, IC3IoT 2018, 2019, pp. 111–116, 8668149

[25] Kala, S., Nandan, H., Sharma, P., Shadow and weak gravitational lensing of a rotating regular black hole in a non-minimally coupled Einstein-Yang-Mills theory in the presence of plasma, European Physical Journal Plusthis link is disabled, 2022, 137(4), 457

[26] Kandpal, H., Kumar, A., Malik, A., Morphometric analysis and prioritization of sub-watersheds in Bino watershed, Uttarakhand: A remote sensing and GIS perspective, Indian Journal of Ecology, 2018, 45(3), pp. 475–478

[27] Kannan, M., Singh, S., Prasad, R.R., Synthetic Methods to Obtain Calcia-Stabilized Zirconia Powders—A Review, Lecture Notes in Mechanical Engineeringthis link is disabled, 2021, pp. 405–416

[28] Kukreti, A., Painolli, A.K., Rana, N., Where do we stand: factors affecting sustainable development, Procedia Environmental Science, Engineering and Management, Volume 8, Issue 2, 2021, Pages 583-589

[29] Kulkarni, S & Chima, P 2021 'Challenges faced by UK university students due to thcoronavirus crisis in the Higher Education' Preprints.org. https://doi.org/10.20944/preprints202102.0192.v1

[30] Kulkarni, S & Sodre, JR 2017, An Assessment of Lightweight Vehicles to Reduce Greenhouse Gas Emissions with Focus on Magnesium. in Thoughts and Refections on the use of Magnesium. Birmingham City University, pp. 84-91.

[31] Kulkarni, S, Chapman, C & Shah, H 2016 'Computational Fluid Dynamics (CFD) Mesh Independency Study of A Straight Blade Horizontal Axis Tidal Turbine' Preprints.org. https://doi.org/10.20944/preprints201608.0008.v1

[32] Kulkarni, S, Chapman, C, Shah, H & Edwards, D 2018, 'A computational design method for bio-mimicked horizontal axis tidal turbines', International Journal of Building Pathology and Adaptation, vol. 36, no. 2, pp. 188-209. https://doi.org/10.1108/IJBPA-06-2017-0029

[33] Kulkarni, S, Chapman, C, Shah, H, Parn, E & Edwards, DJ 2018, 'Designing an efficient tidal turbine blade through bio-mimicry: A systematic review', Journal of Engineering, Design and Technology, vol. 16, no. 1, pp. 101-124. https://doi.org/10.1108/JEDT-08-2017-0077

[34] Kulkarni, S, Chapman, C, Shah, H, Parn, E & Edwards, DJ 2018, 'Design Study of Horizontal Axis Tidal Turbine Blade', Mindanao Journal of Science and Technology. https://www.open-access.bcu.ac.uk/10089/

[35] Kulkarni, S, Edwards, DJ, Chapman, C, Hosseini, MR & Owusu-Manu, D-G 2019, 'A preliminary mechanical design evaluation of the Wikispeed car: for light-weighting implications', Journal of Engineering, Design and Technology, vol. 17, no. 1, pp. 230-249. https://doi.org/10.1108/JEDT-09-2018-0154

[36] Kulkarni, S, Edwards, DJ, Parn, E, Chapman, C, Aigbavboa, CO & Cornish, R 2018, 'Evaluation Of Vehicle Lightweighting To Reduce Greenhouse Gas Emissions With Focus On Magnesium Substitution', Journal of Engineering, Design and Technology, vol. 16, no. 6, pp. 869-888. https://doi.org/10.1108/JEDT-03-2018-0042

[37] Kulkarni, SS, Wang, L, Golsby, N & Lander, M 2021, 'Fluid-structure interaction based optimisation in tidal turbines: A perspective review', Journal of Ocean Engineering and Science. https://doi.org/10.1016/j.joes.2021.09.017

[38] Malik, A., Kumar, A., Kandpal, H., Morphometric analysis and prioritization of sub-watersheds in a hilly watershed using weighted sum approach, Arabian Journal of Geosciencesthis link is disabled, 2019, 12(4), 118

[39] Mohanakurup, V.K., Baghela, V.S., Kumar, S., Srivastava, P.K., Doohan, N.V., Soni, M., Awa, H., 5G Cognitive Radio Networks Using Reliable Hybrid Deep Learning Based on Spectrum Sensing, Green Internet of Things in 5G and Beyond, Hindawi, Volume 2022

[40] Mohta, M., Soni, N., Choudhury, T., Kumar, V., Market Analysis on Li-Fi Technology, Proceedings of the 2019 8th International Conference on System Modeling and Advancement in Research Trends, SMART 2019, 2020, pp. 139–143, 9117427

[41] Nagwan, H., Rana, N., Kumar, A., Tewari, P.C., Performance and emission characteristics of diesel engine using Karanja oil methyl ester and Eucalyptus oil fuel blends, Procedia Environmental Science, Engineering and Management, Volume 6, Issue 4, 2019, Pages 553-563

[42] Painoli, A.K., Bansal, R., Singh, R., Kukreti, A., Impact of digital marketing on the buying behavior of youth with special reference to Uttarakhand state, Big Data Analytics for Improved Accuracy, Efficiency, and Decision Making in Digital Marketing, 2021, pp. 162–182

[43] Parag, J. (2011). Hybrid mixed handover for call blocking probability analysis in wireless ATM network. Journal of Engineering and Computer Innovations, 2(4), 68-77.

[44] Sharma, C., Singh, M.P., Chaudhary, S., Cost Analysis of Solar-Wind Based Hybrid Renewable Energy System, Lecture Notes in Electrical Engineeringthis link is disabled, 2021, 700, pp. 2733–2746

[45] Sharma, S., Chandra, U., & Jain, P. (2017). A literature survey on automation of test data generation for branch coverage testing using genetic algorithm. International Journal of Computational Intelligence Research, 13(6), 1521-1531.

[46] Varshney, R., Jain, P., & Vijay, S. (2018, September). Massive MIMO Systems In Wireless Communication. In 2018 2nd International Conference on Micro-Electronics and Telecommunication Engineering (ICMETE) (pp. 39-44). IEEE.

CHAPTER - 02

Augmented Reality Framework for Distributed Collaboration Using Machine Learning

Dr. Anil W. Kale

Mahatma Gandhi Mission's College of Engineering and Technology, India.
Email: anil5474@gmail.com

Dr. Nandkishor P Karlekar

MGMCET Kamothe Navimumbai, India. Email: npkarlekar.mgmcet@gmail.com

Dr. Vaibhav Narawade

Ramrao Adik Institute of Technology, India. Email:vnarawade@gmail.com

Dr. Sanjay B. Waykar

Mahatma Gandhi Mission's College of Engineering and Technology, India.
Email: sbwaykar@gmail.com

INTRODUCTION

However, lightweight and extensible, powerful platforms are also need powerful frameworks that are possible to create more advanced Virtual and Augmented Reality (VAR) frameworks (software that leverage both VAR and AR). Virtual and augmented reality has been experimented with in a number of different equipment and network settings in the background. They all follow a similar overall structure, but each is different.

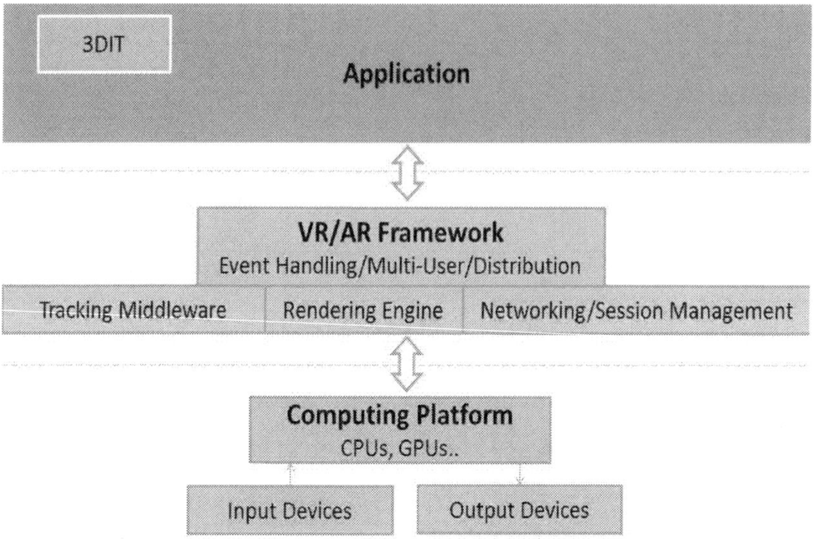

Fig. 1: Virtual reality and augmented reality network design.

The key components of an AR/VR 's system in figure 1 as being an extensive two inputs devices and inclination sensors, a capable of tracking, and a programme running an Analytic framework on a potent cloud infrastructure equipped with a GPU (or graphics processing unit) and an AR system and comprehensive AR structure Most importantly, there is one (or several) user (or a group of users) who is involved from a certain activity and thcy are communicating with the machine.

Towards complete body collection are used to monitor proposed several solutions, sensors, and movement of various body sections of the user and/devices; to measure upstream and downstream movements; movement of various techniques are used to assess the user's position the computer interface (e.g. desktop pc, cell device) receives data about these machines and expands the data to include it in the programme. Although the analysis delivered to the customer depends on this approach, the data would be expanded to 3D. With recent computer games devices, and more efficient technology being accessible, people may use artificial intelligence to work and social media with partners at the same time, novel connectivity options can be offered at low costs.

Event management handles certain users' input/devices, but the number of input devices (e.g., fingers) and how they're interacting are distributed through different devices (single) users is managed by a system (layer). Later, the scene is rendered out on the display computer such that the simulated world can be shown to the viewer. Ideally, ideally, the headset will provide high-quality rendering, support for mechanics, and genuine connectivity and scenario control could be offered as a component of a virtual/augmented reality (AR) technology development environment. Consequently, a virtual reality tools must be simple to learn, cheap, portable, and compatible with the various available equipment, and capable of filling up and easy for testing and educational use, while still having accurate documentation and accurate simulation to meet the needs of training.

Some issues outlined in section II also need to be addressed, such as the price, versatility, and stretchability of emerging toolkits and methods. As there was no need for a bespoke virtual reality or augmented reality software development, we created a platform built on an off-the-the-shelf engine for serving several users with a diverse set of input technologies. All in all, our architecture has the following capabilities: a virtual reality/augmented reality framework that can be quickly developed using a word processor and control of scenes. It can serve as an infrastructure for VR/AR/AR solutions that are highly flexible in both platform and deployment configuration, and compatible with both on-the-site and off-site and, in addition platforms allow you to run VR/deploy different settings with numerous touch screens such as 2D mice, VR motion markers, and highly correlated and projections, as well as 6. Besides this, you can run VR and AR applications on several software platforms as well as on smart phones, like HoloLens, in VR headsets and in augmented reality, it is also supported by 3D mice, virtual reality controls, and AR devices such as the Kinect and ODG based on the viewing glasses (HMD).

Play Station, we propose the following 3 primary achievements: (1) establish an interactive platform for the multi- as well as the on-desk VR and AR setups, (2) full and open tool support for brand-new management software options, and (3) fully, less expensive controls like Windows Move, Razer Hydraviz, and Sony Move we propose these: (3) (2) including cheap choices including Hydraviz and Google and Sony PlayStation, we've designed and then built up an open system with minimal additional requirements for both brand-new and legacy operating systems.

However, in Augmented reality, user himself is augmented in the scene along with virtual objects in an immersive way such that user feels he is directly interacting with objects. His movement is resolved through sensory analysis and is transformed to characteristic Overall methodology is divided into above depicted conceptual blocks. Our work should focus towards developing services specific to each of the mentioned components such that they can be utilized by broader range of devices. As mobiles and other hand help devices lags the power of personal computers while outgrowing them in usage, cloud based services address this issue by extending the augmented reality fundamentals to be extended to devices with less computational power.

Study of Relevance

A multitude of VR technologies have been established and widely since the mid-1990s a vast range of VR architectures have appeared, as well as virtualized resources serving wide range of applications Many architectures are built on scene diagram modules, for example, open source examples such as VR Thrower or Virtools, as well as proven products like 3DIA Virtools. The help includes many input and output equipment may be varied in size. That after this, we shall briefly explore our structures that are relevant to our job.

A platform for wearable technology (AR) users to write code and projects together. As of 2018, it's been about a decade since it was established, construction began in 1996 and lasted for around ten years for academic and scientific purposes. While several users can interact with the asset, it, it is capable of managing a virtual environment of programs which are placed as nodes within the scene graph. Conventional proprietary image editing applications tend to be complex, expensive to develop and impossible to use. One of the reasons is the inability to provide users a visual representation of the process, because of the presence of image treatment modalities that are less open. Also more modern technology, such as cell phones or sonar sensors, may be almost impossible to incorporate into the surveys, for that purpose.

It's a third-party service portal that helps immersive 3D content creators create and distribute their 3D assets. using manufacturing VR adapters, it enables the use of several users while still objects and the laws of physics to generate virtual environments, which make it possible for interactive and decentralized experiences You will have a lot more control and flexibility over your performance, whether you create in graphics or text-mode, with the result that you can provide a broad variety of scope of this article with it. Nevertheless, it's prohibitively expensive for use in the scientific and educational communities, since it is out of reach for most people.

FRAMEWORK OVERVIEW & DATA FLOW

As per our inspiration in section above, we wanted to build a scalable software that is compatible with new devices and interaction methods, we proceeded to create an expanded architecture that is adaptive to each specific hardware configuration. a look at the system and how it is built as well as all of its elements and data flow. A multi-platform flexible web application layer records workflow structure identifies different inputs, from various fileserver computers as well as from portable apps, and utilises these characteristics to be both consistent and scalable. The platform provides reports to the framework layer in a uniform fashion that is accurate and free of unnecessary variation. An computer on the "expansion pack" on which the programme is designed uses additional modules. starts the implementation layer, when being visualised is served and supplied with a procedure and delivery module that is real-time. ART will manage the raw input details and offer integration and service as well as technical assistance with direct conversations.

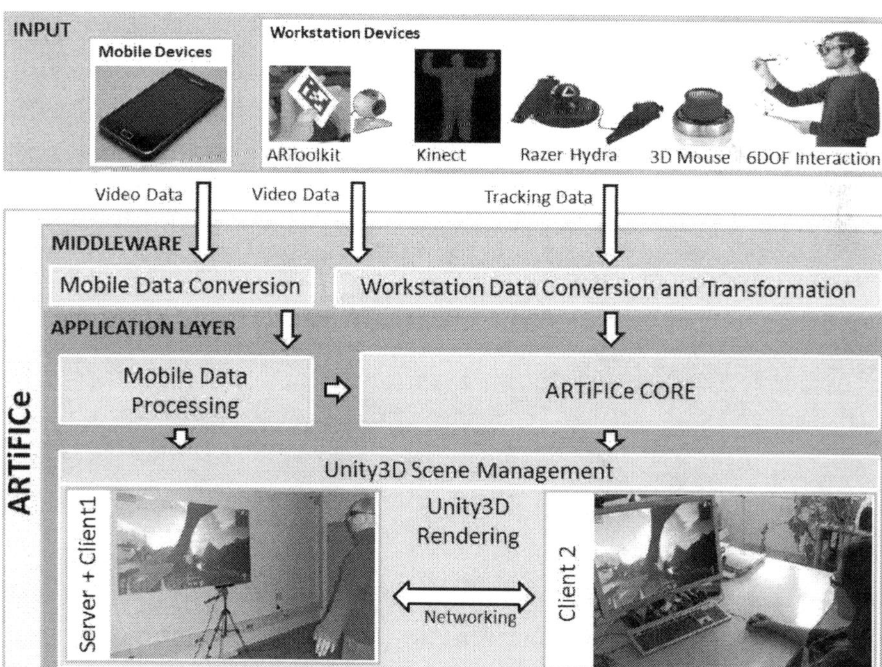

Fig. 2: Framework components and data flow.

The ARTi core features an intuitive architecture that is open to using many two-dimensional and three-dimensional interface design models. It is made modular such that it can incorporate additional design concepts without disrupting how it is organized and to begin with, and provides several modules that is known and isomorphism alternatives to operate on. Via a common link, you can monitor the activity from a workbench or a computer to manage the selection and manipulation of visual features on a virtual computer. Thus, additional people have the ability to engage with 2D and 3D material of the same material in the same area at once leads to plenty of innovative and productive interactions. So we set up a deployment system in the ART core to support multi-user communication with different hardware configurations over a network, allowing the usage of the time-dependent aspect of several users simultaneously. The rest of the relevant information regarding the ARTiF centre are presented in the following sections.

METHODOLOGY

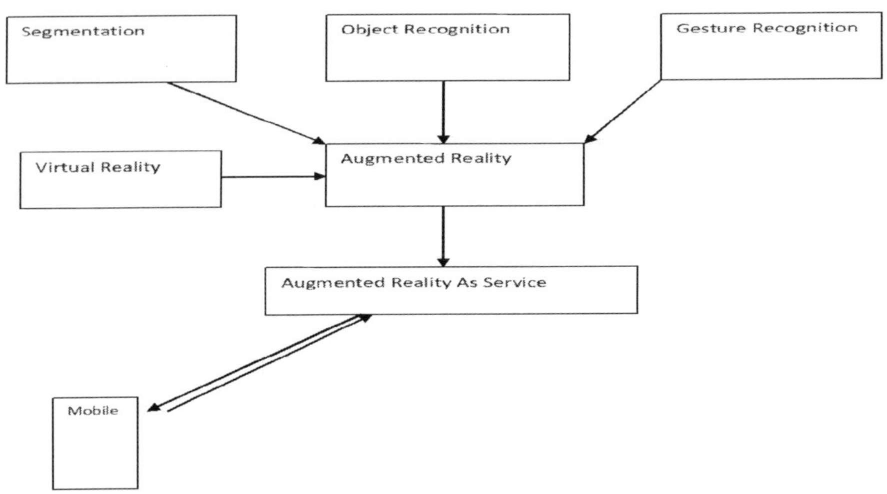

Fig. 3: Methodology

Proposed methodology should first focus towards building a framework for each of this basic building blocks and test their efficiency. The last phase of the research must focus entirely on cloud based extension of the framework followed by a real time usage depiction of the same. Proposed work should be able to integrate derived heuristics in a more holistic way such that the new methods and techniques can seamlessly be used with current standards rather than evoking the demand for all new standards or paradigm shift in them.

POSSIBLE OUTCOME

In the proposed work we mainly focus on building reusable state of art for augmented reality. Firstly, the research would focus towards bettering each of the current state of art building blocks. The methodology must provide techniques that addresses current issues and challenges in augmented reality and offer a more robust and efficient solution for them. One of the first expected result would be a unique framework of Augmented reality where one or more building blocks are texted either in isolation or in conjunction to be meeting real performance objectives of augmented reality.

Then the framework would be used to develop real time application which shows the usability of the developed framework through applications like AR games or novel graphics rendering.

In the last phase, the developed framework and its components would be released as service over cloud which can be consumed by variety of applications. Focus would be to show the applicability of the AR over cloud through mobile based application. Oncc these goals are met, a comprehensive performance analysis would be presented with current technique.

IMPLEMENTATION

Since we want to build a VR and AR platform, we want to use Prompting the government as our network for the whole protocol stack. We are excited about its potent operating system, and its potential to grow, as well as the User base which makes the capabilities of Preach easier to enhance. Additionally, we can take advantage of the license which contains all the tools we need for compiling, encoding, and for optics as well. The use of C++ and C# in both versions is as far as I know.

Web Framework that Supports Middle-tier Networking

Using the package called Open Tracker (found on the system), we collect all input data from different sensors and input method into a single layer for better accuracy. free software instrumentation can be used to expand all types of instrumentation and to be clear and extensible; the two particular properties it possesses are opaque and scalable interface incorporation theoretically, virtual and augmented reality implementations for the various system operations that they choose to implement can use the track voice recognition and the ability to use them in various manners is also essential for supporting.

With this style, objects are used in an entity manner that leverages XML, setup, and metadata resources, the company has control of the whole designs without implementing them in code. While the straightforward model handles linear output, the expandable version takes care of variable tuning and filtering and transforming may be used for monitoring. A lot of commonly used setup devices may be defined by XML files, which would sometimes include a mix of various input devices. Can also be used to access data from some external input systems. Virtual World Private Platform (VRPN) fetches monitoring data from various phones. The Virtual Desktop can be used and networked with almost every laptop in a VR/V setup.

ART incorporates an application-to-layer framework in that it allows for a simple interface with simple layer independence and separation from the data that drives physical systems to the actual state of those devices, thereby improving the functionality of the user experience. To assist in this process, we thereby introduced a new keyword named "sink" for all tools that should be tracked. As stated in section Standalone Software retrieving: ART uses the Unity node as a query object to gather data on-related information in the game logic. On the Open Transaction system, we use VPN (virtual private network) implementations to find and analyze improving transparency.

Predictive surveillance equipment and incorporation. In our platform, we added information for specific control system and allows various setup configurations ranging from desktops to tablet-based, moderately, and full-immersion to virtual reality (VR) to virtual. If you have a laptop software and usage of your own, so you can have convenient access to both the ARTool and ARTkit APIs. These include planar shapes and integrated patterns to assist in monitoring. they are real-time motion compensated, meaning they measure their coordinates related to verbal prompts, location-based sensors and thus allow a broad variety of AR implementations Using the ARTool repository typically happens in a desktop AR application, while adding to it with ARTool+) improves the ARTkit (or is designed for mobile phones) Prior to this, our application, we used a library called ARTool+ that has been group pioneers into.

The image acquisition and analysis system called Open Video is used to receive the scene information from monitor, which is conveyed to Unity3D and ARToolKit+ units, which each expand the details to create a new image that are later streaming into Unity3D for rendering.

We used a Rog Hydra and a 3D mice sensor in the 6-DOF Nodes to enable screen interactions with the cursor and other peripherals. augmented reality systems are best for semi-imaging and semi-active or inactive infrared detection and for ultrasonic level Marker Physical or Passive IR tracking is appropriate in these scenarios.

The machine will automatically detect motion and supply parameters with a tracking rate of 60 Hz, zero position jitter, and a precision of 20-5mm or less, and trace render random physical artefacts using reduced markings with a submillimetre positioning that can achieve a key message less than 0.05mm error. In order to integrate, we first configured the attacker with the OT Virtual Ring with the Unification server.

We decided to expand the marker live actors so we could map not just the markers but the whole bodies as well. In order to get Kinect as a part of the system, we used the OpenNI /FAAST plugin. An openni expand/NITE exposes raw depth details and the detail-based skeleton data, with the benefit of applying additional detail processing. FAAST operates as a body programmed and is able to accesses this info. gesture help and complete body motion detection details are also included in the product as an Expansion pack for the Oculus Touch We are feeding the ART backbone and real-node real-time monitoring through the Expansion Node to keep the Expansion in sync.

In order for these new mobile platforms to take advantage of both augmented reality and virtual reality technologies, there must be a proper virtual reality (AR) platform architecture behind them. Developing android Mobile apps for the use of the monitor as a "magic lens" as a tool is made possible by making Vuforia function for the user. With 3D perception, it's sensor hands over a range of different features such as rigorous monitoring and pose to the augmented reality software, which lets you use it to create exclusive 3D models and delivers it to the monitor so you can view live footage in your screen. This app is made for iOS and Android users. Vuforia is wachter for both data-hand lingered devices and mobile devices, which in our case functions works as a mobile part connected to the ARTiCore. We have a modular middleware definition that lets us control all the machines in a simple XML admin panel, or in several different files utilising a consistent platform. Configuring mobile devices uses Vuforia is separate from the user experience.

It implements a modular and effective game engine-based development architecture that is used to create shared and interactive virtual and applied projects for study and teaching. To have a wide range of 6-degree of freedom of motion for our virtual environments, we implemented lower cost Microsoft Kinect, Hydra, and Razer's Space Explorer 6-based devices into our architecture. One of the major goals of augmented reality is creating an environment where user is immersed in the scene along with other objects. In a virtual reality, user deals with only virtual objects being rendered in an artificial scene. Augmented reality is the process of rendering virtual objects in a scene such that the scene appears as mixed reality. Augmented Reality based health applications are gaining popularity due to improvement of processing and rendering techniques and open source software. Several cameras based health vital extraction techniques are being developed over last several years. However, these techniques are limited time applications, and are limited by the lighting, environment and posture. In this work we propose a unique cloud based augmented reality health vital framework that extracts continuous pulse rate and pulse signal from user face image using cloud based face detection, intensity normalization, facial contour tracking, atrial vascular network correlation on face and temporal peak detection. The proposed method is validated in real time under controlled clinical environment and provides an overall accuracy of 91%. The system also augments continuous health information on the real camera feed.

RESULTS

We used various configurations to see how well our architecture worked and to deal with varying workloads. At the time of this writing, much of the system is still running on Platforms (with the exception of a few sections, such as Kinect and ART about which there is some Linux support for some aspects of it. ARTiFlex is additionally used to build Android apps (as well as new Android platforms) For eighth graders, the team designed four Virtual Reality (VR) and Augmented Reality (AR) playlets by using ARTool Kit's visual programming tools. After the use of the different methods and technologies from Connexion Space Navigator and Windows Kinect, teachers and parents worked together to create a scalable and interactive 3D/2D/AR VR/AR app.

Furthermore, our system was also utilized for the Laboratories upper Austria for the fall semester in 2011/2012 academic year. The students should be able to quickly and efficiently create a smartphone shared and shared virtual/augmented reality (VR/AR) technology for 17 people in four weeks. and Hydra, 6DOF desktop has already found use in many on-going study projects in which the experience is tied directly to the physical interface environment, where any individual and device interacts as if the usage of a VR headsets.

We specified the film and tv possibly the best inputs and loose linking of instructional strategies, allowing more advanced technicians to adjust to their particular needs. creation of a framework in comparison, we offer an effective platform to help students bypass the more challenging parts of building simple AR experiences by using prototyping to quickly jump through their first obstacles.

REFERENCES

[1] A. Forsberg, K. Herndon and R. Zeleznik, Aperture based selection for immersive virtual environments, Proceedings of the 9th ACM symposium on user interface software & technology, pp. 95-96, 1996

[2] Abhigna, B.S., Chakraborty, S., Kumar, V., Medha: A collaborative E-concept framework, Proceedings of the 2018 International Conference On Communication, Computing and Internet of Things, IC3IoT 2018, 2019, pp. 117–122, 8668127

[3] Abidin, S., Vadi, V.K., Rana, A., On Confidentiality, Integrity, Authenticity, and Freshness (CIAF) in WSN, Advances in Computer, Communication and Computational Sciences, 2020, pp 87-97

[4] Ahay, A., Choudhury, T., Kumar, V., Hierarchical Evolution of Digital Games, Proceedings of 2019 International Conference on Computational Intelligence and Knowledge Economy, ICCIKE 2019, 2019, pp. 381–386, 9004378

[5] AIAA Aerospace Sciences Meeting and Exhibit Reno, USA, 2002

[6] Area: Communication, Frequency modulation, Networks, wireless communication

[7] Area: Communication, Frequency modulation, Networks, wireless communication

[8] Area: Communication, Frequency modulation, Networks, wireless communication

[9] Area: Communication, Frequency modulation, Networks, wireless communication

[10] Arora, M., Kumar, V., Dave, M., Hybrid cloud infrastructure for load balancing and scaling with vm path optimization using ant colony optimization, Journal of Advanced Research in Dynamical and Control Systems, 2020, 12(5 Special Issue), pp. 660–674

[11] Arora, M., Kumar, V., Dave, M., Task scheduling in cloud infrastructure using optimization technique genetic algorithm, Proceedings of the World Conference on Smart Trends in Systems, Security and Sustainability, WS4 2020, 2020, pp. 788–793, 9210303

[12] Bansal, R., Gupta, A., Singh, R., Nassa, V.K., Role and impact of digital technologies in E-learning amidst COVID-19 pandemic, Proceedings - 2021 4th International Conference on Computational Intelligence and Communication Technologies, CCICT 2021, 2021, pp. 194–202

[13] Bansal, R., Obaid, A.J., Gupta, A., Singh, R., Pramanik, S., Impact of Big Data on Digital Transformation in 5G Era, Journal of Physics: Conference Seriesthis link is disabled, 2021, 1963(1), 012170

[14] C. Cruz-Neira, A. Bierbaum, P. Hartling, C. Just, and K. Meinert, VR Juggler ± An Open Source Platform for Virtual Reality Applications, 40th

[15] Chauhan, S.S., Khare, A.K., Kinematic analysis of the ABB IRB 1520 industrial robot using roboanalyzer software, Evergreenthis link is disabled, 2020, 7(4), pp. 510–518

[16] Chauhan, S.S., Khare, A.K., Method of Obtaining Planar State of Stress Using Mohr's Circle—Some Typical Cases, Lecture Notes in Mechanical Engineering, 2022, pp. 33–38

[17] D. Mizen. Augmented reality applications in aerospace. In Proceedings of ISAR, 2000.

[18] D. Nicklas and B. Mitschang. A model-based, open architecture for mobile, spatially aware applications. In Proceedings of Advances in Spatial and Temporal Databases, pages 392–401, 2001.

[19] D. Schmalstieg, G. Schall, and D. Wagner. Managing complex augmented reality models. IEEE Computer Graphics and Applications, 27(4):48–57, 2007.

[20] D.A. Bowman and L.F. Hodges, An evaluation of techniques for grabbing and manipulating remote objects in immersive virtual environments, Proceedings of the 1997 ACM symposium on Interactive 3D graphics, pp. 35ff, 1997

[21] G. Hesina, Distributed collaborative augmented reality, PhD thesis, Vienna University of Technology, 2001

[22] G. Reitmayr and D. Schmalstieg. Collaborative augmented reality for outdoor navigation and information browsing. In Proceedings of ymposium Location Based Services and Tele Cartography, pages 53–62, 2004.

[23] Goel, S., Tripathi, A., Choudhury, T., Kumar, V. Parkinson's Disease Detection using Soft Computing Technique, Proceedings of the 2019 8th International Conference on System Modeling and Advancement in Research Trends, SMART 2019, 2020, pp. 3–6, 9117384

[24] Govindarajan, V., Sivakumar, R., Patil, P.P., Kannan, M., Ramesh, B., Effect of Tungsten Carbide Addition on the Microstructure and Mechanical Behavior of Titanium Matrix Developed by Powder Metallurgy Route, Advances in Materials Science and Engineeringthis link is disabled, 2022, 2022, 2266951

[25] Gupta, A., Singh, R., Nassa, V.K., Sharma, P., Koti, K., Investigating Application and Challenges of Big Data Analytics with Clustering, 2021 International Conference on Advancements in Electrical, Electronics, Communication, Computing and Automation, ICAECA 2021, 2021

[26] H. Kato, M. Billinghurst, R. Blanding, and R. May. Artoolkit, 1999.

[27] J. Irizarry, M. Gheisari, and G. Williams. Infospot: A mobile augmented reality method for accessing building information through a situation awareness approach. J. Autom. Constr., 33:11–23, 2012.

[28] Jain, P., Vijay, S., & Gupta, S. C. (2009). ATM Network QoS Parameters and Performance Analysis using OPNET Simulator. International Journal of Recent Trends in Engineering, 2(4), 228.

[29] Jain, P., Vijay, S., & Gupta, S. C. Performance Analysis & QoS Guarantee in ATM Networks. Global Journal of computer Science and Technology, 131-136.

[30] Jain, S., Choudhury, T., Kumar, V., Kumar, P., Detecting malware and analysing using sandbox evasion, Proceedings of the 2018 International Conference On Communication, Computing and Internet of Things, IC3IoT 2018, 2019, pp. 111–116, 8668149

[31] Kala, S., Nandan, H., Sharma, P., Shadow and weak gravitational lensing of a rotating regular black hole in a non-minimally coupled Einstein-Yang-Mills theory in the presence of plasma, European Physical Journal Plusthis link is disabled, 2022, 137(4), 457

[32] Kandpal, H., Kumar, A., Malik, A., Morphometric analysis and prioritization of sub-watersheds in Bino watershed, Uttarakhand: A remote sensing and GIS perspective, Indian Journal of Ecology, 2018, 45(3), pp. 475–478

[33] Kannan, M., Singh, S., Prasad, R.R., Synthetic Methods to Obtain Calcia-Stabilized Zirconia Powders—A Review, Lecture Notes in Mechanical Engineeringthis link is disabled, 2021, pp. 405–416

[34] Kukreti, A., Painolli, A.K., Rana, N., Where do we stand: factors affecting sustainable development, Procedia Environmental Science, Engineering and Management, Volume 8, Issue 2, 2021, Pages 583-589

[35] Kulkarni, S & Chima, P 2021 'Challenges faced by UK university students due to thcoronavirus crisis in the Higher Education' Preprints.org. https://doi.org/10.20944/preprints202102.0192.v1

[36] Kulkarni, S & Sodre, JR 2017, An Assessment of Lightweight Vehicles to Reduce Greenhouse Gas Emissions with Focus on Magnesium. in Thoughts and Refections on the use of Magnesium. Birmingham City University, pp. 84-91.

[37] Kulkarni, S, Chapman, C & Shah, H 2016 'Computational Fluid Dynamics (CFD) Mesh Independency Study of A Straight Blade Horizontal Axis Tidal Turbine' Preprints.org. https://doi.org/10.20944/preprints201608.0008.v1

[38] Kulkarni, S, Chapman, C, Shah, H & Edwards, D 2018, 'A computational design method for bio-mimicked horizontal axis tidal turbines', International Journal of Building Pathology and Adaptation, vol. 36, no. 2, pp. 188-209. https://doi.org/10.1108/IJBPA-06-2017-0029

[39] Kulkarni, S, Chapman, C, Shah, H, Parn, E & Edwards, DJ 2018, 'Designing an efficient tidal turbine blade through bio-mimicry: A systematic review', Journal of Engineering, Design and Technology, vol. 16, no. 1, pp. 101-124. https://doi.org/10.1108/JEDT-08-2017-0077

[40] Kulkarni, S, Chapman, C, Shah, H, Parn, E & Edwards, DJ 2018, 'Design Study of Horizontal Axis Tidal Turbine Blade', Mindanao Journal of Science and Technology. https://www.open-access.bcu.ac.uk/10089/

[41] Kulkarni, S, Edwards, DJ, Chapman, C, Hosseini, MR & Owusu-Manu, D-G 2019, 'A preliminary mechanical design evaluation of the Wikispeed car: for light-weighting implications', Journal of Engineering, Design and Technology, vol. 17, no. 1, pp. 230-249. https://doi.org/10.1108/JEDT-09-2018-0154

[42] Kulkarni, S, Edwards, DJ, Parn, E, Chapman, C, Aigbavboa, CO & Cornish, R 2018, 'Evaluation Of Vehicle Lightweighting To Reduce Greenhouse Gas Emissions With Focus On Magnesium Substitution', Journal of Engineering, Design and Technology, vol. 16, no. 6, pp. 869-888. https://doi.org/10.1108/JEDT-03-2018-0042

[43] Kulkarni, SS, Wang, L, Golsby, N & Lander, M 2021, 'Fluid-structure interaction based optimisation in tidal turbines: A perspective review', Journal of Ocean Engineering and Science. https://doi.org/10.1016/j.joes.2021.09.017

[44] M. Billinghurst, M., H. Kato, H. and I. Poupyrev, The Magic Book: a transitional AR interface, Computers & Graphics, vol.25:5, pp. 745-753, Elsevier, 2001

[45] M. Tonnis. Data management for augmented reality applications. Master's thesis, Technische Universitat Munchen, 2003.

[46] Malik, A., Kumar, A., Kandpal, H., Morphometric analysis and prioritization of sub-watersheds in a hilly watershed using weighted sum approach, Arabian Journal of Geosciencesthis link is disabled, 2019, 12(4), 118

[47] Metaio. http://www.metaio.com/, 2014.

[48] Mohanakurup, V.K., Baghela, V.S., Kumar, S., Srivastava, P.K., Doohan, N.V., Soni, M., Awa, H., 5G Cognitive Radio Networks Using Reliable Hybrid Deep Learning Based on Spectrum Sensing, Green Internet of Things in 5G and Beyond, Hindawi, Volume 2022

[49] Mohta, M., Soni, N., Choudhury, T., Kumar, V., Market Analysis on Li-Fi Technology, Proceedings of the 2019 8th International Conference on System Modeling and Advancement in Research Trends, SMART 2019, 2020, pp. 139–143, 9117427

[50] Nagwan, H., Rana, N., Kumar, A., Tewari, P.C., Performance and emission characteristics of diesel engine using Karanja oil methyl ester and Eucalyptus oil fuel blends, Procedia Environmental Science, Engineering and Management, Volume 6, Issue 4, 2019, Pages 553-563

[51] Painoli, A.K., Bansal, R., Singh, R., Kukreti, A., Impact of digital marketing on the buying behavior of youth with special reference to Uttarakhand state, Big Data Analytics for Improved Accuracy, Efficiency, and Decision Making in Digital Marketing, 2021, pp. 162–182

[52] Parag, J. (2011). Hybrid mixed handover for call blocking probability analysis in wireless ATM network. Journal of Engineering and Computer Innovations, 2(4), 68-77.

[53] R. Freitas and P. Campos. Smart: a system of augmented reality for teaching 2nd grade students. In Proceedings of British Computer Society Conference on Human-Computer Interaction, pages 27–30, 2008.

[54] S. J. Henderson and S. K. Feigner. Augmented reality in the psychomotor phase of a procedural task. In Proceedings of ISMAR, pages 191–200, 2011.

[55] S. Julier, Y. Baillot, L. M, and D. Brown. Bars: Battlefield augmented reality system. In NATO Information Systems Technology Panel Symposium on New Information Processing Techniques for Military Systems, 2000.

[56] S. Kosta, A. Aucinas, and P. Hui. Thinkair: Dynamic resource allocation and parallel execution in cloud for mobile code offloading. In Proceedings of INFOCOM, pages 945–953, 2012.

[57] Sharma, C., Singh, M.P., Chaudhary, S., Cost Analysis of Solar-Wind Based Hybrid Renewable Energy System, Lecture Notes in Electrical Engineeringthis link is disabled, 2021, 700, pp. 2733–2746

[58] Sharma, S., Chandra, U., & Jain, P. (2017). A literature survey on automation of test data generation for branch coverage testing using genetic algorithm. International Journal of Computational Intelligence Research, 13(6), 1521-1531.

[59] T. Hollerer, S. Feiner, and D. Hallaway. User interface management techniques for collaborative mobile augmented reality, computers and graphics. Modeling and Simulation Design, 25(5):799–810, 2001.

[60] Varshney, R., Jain, P., & Vijay, S. (2018, September). Massive MIMO Systems In Wireless Communication. In 2018 2nd International Conference on Micro-Electronics and Telecommunication Engineering (ICMETE) (pp. 39-44). IEEE.

[61] W. Holden. http://www.juniperresearch.com/viewpressrelease.php pr=166, 2009.

[62] W. Holden. http://www.juniperresearch.com/viewpressrelease.php pr=232, 2011.

[63] W. Piekarski and B. H. Thomas. Arquake: The outdoor augmented reality gaming system. Communications of the ACM, 45(1):36–38, 2002.

[64] W. Piekarski and H. T. Bruce. An object-oriented software architecture for 3d mixed reality applications. In Proceedings of ISMAR, pages 247–256, 2003.

[65] Z. P. Huang, P. Hui, C. Peylo, and D. Chatzopoulos. Mobile augmented reality survey: A bottom-up approach. Technical report, HKUST Technical Report, 2013 (cite as arXiv:1309.4413).

CHAPTER - 03

Machine Learning and AI (Health Care)

Dr. Nandkishor P Karlekar

MGMCET Kamothe Navimumbai, India. Email: npkarlekar.mgmcet@gmail.com

Dr. Vaibhav Narawade

Ramrao Adik Institute of Technology, India. Email:vnarawade@gmail.com

Dr. Sanjay B. Waykar

Mahatma Gandhi Mission's College of Engineering and Technology, India. Email:sbwaykar@gmail.com

Dr. Anil W. Kale

Mahatma Gandhi Mission's College of Engineering and Technology, India. Email: anil5474@gmail.com

INTRODUCTION

Symptomatic instruments further develop the general medical care by conveying basic data to the appropriate people in real conditions. The general objective of powerful medical care is to make it simpler for the patient to get information about their findings and the best strategy for settling them. Through understanding the numerous parts of medical care, patients are engaged to react properly in case of intense disease. It urges medical services experts to broaden their practices by allowing a more worldwide far off turn upward to guarantee investment funds and empowers more savvy therapies for their patients through permitting the utilization of more current mechanical developments; a far reaching medical services foundation is expected to guarantee individuals have the wellbeing therapy they need. Aside from actual wellness, one of the basic objective is expanding medical care uses is to a push ahead by-regardless of step early analysis. As another model, the IoT area, this is anticipated to be worth 158.1 billion out of 2022, individual medical care the executives with the Internet of Things.

The Internet of Things will prompt a progress in medical care and lower the expense of clinical gear. IoT will acquire considerable use because of the sending would happen on account of 5G organizations. 5G organizations are set to be progressively well known for emergency clinics, clinical offices, and medical care suppliers. This is the fifth influx of versatile organization (5G). There are three ages of organizations following the primary 2G, 3G, 3G, and 4G organization. Another sort of organization has been fostered that interfaces basically all and everything to each other, making it conceivable to consolidate the mechanical and the actual domains. Regularly called multi Gig hertz remote innovation, which is promoted as having the option to give the Gbps, ultra-hertz transfer speed, colossal force, and consistency for customers, just as generous and generally development in speed for the organization utilization for clients. Expanded creation and expanded usefulness empower new industry areas to do what they already couldn't and help them to do all alone.

Applications that incorporate availability like public security and government utilities, and transportation, just as brilliant gadget portability for clinical and wellbeing checking, and enormous information are on 5G through and the colossal IoT. Albeit 5G is worked to have the ability to give different organizations over the long haul, it should be extended to be made adaptable to oblige the administrations that we don't yet perceive.

Upgraded Portable Broadband

As well as making our more brilliant telephones more productive, 5G will empower an expansion, all things considered, inundation encounters like VR and AR, higher information speeds, and less-per-second execution, it will take into account a decline in dormancy.

Those Fundamental for Achieving the Mission

No new help has arisen as an industry relies so vigorously upon high on super solid, broadly open, low-idleness associations as doing as such with the 5G-empowered arrangement.

Quickly Growing Internet of Things

The adaptability and versatility of 5G innovation was expected to assist with diminishing information speeds, asset use, and openness costs for almost whatever is introduced in a wide scope of articles in the organization, taking into consideration exceptionally productive and adaptable, however savvy, arrangements. With the multiplication of remote foundation, brilliant regulators, and keen gadgets, various enterprises are depending on disseminated machine-to-machine (M2M) conditions and web of things (IoT) for working on functional productivity.

Existing M2M organizations require various regulators and heterogeneous equipment/correspondence interfaces to interface the M2M sensors to their particular focal units to get measure and send information. In any actual climate, the sent M2M/IoT stage gathers organized and unstructured information and an assortment of related and inconsequential never-ending occasions which are handled by at least one hubs relying upon its physical and useful capacities.

The regulators and interfacing M2M gadgets generally have pre-taken care of programming that takes into account determined circumstances and where the runtime works on a proper arrangement of rules/rationale inside its limit of tasks. Because of the imperatives of preparing and correspondence limit, any single hub with preloaded underlying knowledge will have restricted extension to associate and total crude occasions, create situational mindfulness and make brief moves to deal with the problematic functional climate of a M2M/IoT environment. These frameworks have restricted capacity to use machine knowledge and cooperative learning inside a predefined framework limit. The cognizance interaction in these frameworks is for the most part bound to information mining and in view of restricted pre-taken care of knowledge.

As specialists have done in the course of the most recent couple of years, record and patient's experience data is held for a few ages. With patients who decide to utilize the investigation, the records would have been kept in an office place so they may see it. the issue with communicating clinical notes to the primary second anybody sees a patient to another specialist is that it is a mix-up Once the patient's records had been shipped off the new doctor, which will be accomplished through reflecting and movement. Lamentably, duplicates can't be delivered and at this point don't move, on the grounds that there would be nothing else of the asset.

It has been set up that electronic checking for medical services suppliers and their patients to assist individuals with bettering screen their wellbeing and cling to therapy plans It is regularly when medical services associations apply such predictable practices, are in any event, while moving them to another association, frequently unintentionally, that information might be ported.

Normally, there are a few classified records situated in workplaces that are private, including emergency clinics, army installations, and outsider safety net provider offices that can't be found by the patient's present specialist or advisor. Albeit this data is unquestionably essential to patients, all things considered, it very well might be scattered to the doctors, centers, and everybody in a question of crisis as I in a period of critical clinical scarcity, for I have thirty wellbeing organizations to work with, there isn't an approach to send it to the overall population and to such gatherings as that as effectively as this to the current establishments.

As this assignment of making clinical data all the more broadly accessible has been cultivated, admittance to it will be altogether streamlined. Notwithstanding all the should be distant from everyone else, being overlooked, unfortunately, permitting autonomous individuals the ordinary chance to utilize the innovation, just as protected people to see clinical records is an issue we actually need to confront. Patients may be in a difficult situation when endeavoring to track down their clinical data, on the grounds that the records are generally put away in a facility, and not in every case close by.

However, by and large, the vast majority of the records stay at the center or medical clinic where a patient's fundamental doctor or attendant specialist is found. The people who travel excessively far from their center for treatment is by all accounts unconscious of this material, nonetheless, regularly track down that the greater part of it is on record or not carefully accessible.

OBJECTIVES

1. The target of the innovation is to a doing is moving from a conventional expert and emergency unit a scattered clinic, and there are numerous advances in transit that are empowering this transformation of "from the base up". The machine's innovation starting today, 5G and other systems administration gadgets are in medical care, 5G frameworks are generally being used for getting to medical services offices of the learned person, and patient administration advancements are headed to advance.
2. The other target of the innovation is to a Due to the expected expansion in the volume of information created by present day approaches like modernized medical services, virtual medical care, and other required applications, we ought to expect a

gigantic piles of Exabyte of pet bytes (or Exabyte of gigabytes) scale information to open up in various configurations and measurements. Given the complex and time-delicate wellbeing executions, 5G gadgets need broad finish to-to-end, edge, or highlight to-point delay, and different components, or attributes to be streamlined.

3. The other target of the creation is to a the improvement of new advances Device-explicit upgrades of 5G systems administration being as of now are being acquainted with tackle different wellbeing explicit advances like the Internet of Things (IoT) to give exceptionally customized medicines, while the more extensive sending of savvy networks regularly tries to connect unique populaces like the IoT organizations and IoT gadgets for omnipresent medical services applications that request explicit quality and expanding cell and region infiltration.
4. The other target of the innovation is to a Within the setting of this investigation, we talk about current 5G IoT-related issues like adaptability, productivity, execution, openness, gadget and streamlining, we tackle less energy use, versatility, gadget accessibility, and battery life, alongside more-permitting gadgets to be gotten to, convenience, and insurance are settled. Under AI, we have been attempting to grow the current admonition with the goal that such analytic medical services elements and way of life propensities can be represented and find more proficient future therapies for by investigating data and automatable use situations identified with these elements.

METHOD

Exabyte of pet bytes (or Exabyte of gigabytes) scale information to open up in various organizations and measurements. Given the complex and time-touchy wellbeing executions, 5G gadgets need broad finish to-to-end, edge, or highlight to-point delay, and different components, or attributes to be streamlined. the improvement of new advancements Device-explicit upgrades of 5G systems administration being right now are being acquainted with tackle different wellbeing explicit innovations like the Internet of Things (IoT) to give profoundly customized medicines, while the more extensive arrangement of wise organizations regularly tries to interface unique populaces like the IoT organizations and IoT gadgets for pervasive medical services applications that request explicit quality and expanding cell and region infiltration. Inside the setting of this examination, we talk about current 5G IoT-related issues like versatility, effectiveness, execution, openness, gadget and advancement, we tackle less energy use, adaptability, gadget accessibility, and battery life, alongside more-permitting gadgets to be gotten to, ease of use, and insurance are settled. Under AI, we have been attempting to grow the current admonition so such analytic medical services substances and way of life propensities can be represented and find more productive future therapies for by exploring data and automatable use situations identified with these elements.

SUMMARY

A short synopsis of parts and explicit subtleties of the capacity to follow: (A primer definition or not many of many complete ideas canvassed in the extensive extra text) the text is organized so as to exclude a serious investigation of anything in the psyche, however rather centers around different objectives and thought processes. The ensuing audit isn't planned to contain something past of the scope of examination, nor is it expected to have any sort of effect.

As indicated by an excellent element, the straightforwardness, the structure is described by a capacity for substantial and constant or close fundamental choice help for incorporated M2M and IoT applications. The intellectual insight stage has a one of a kind equipment, programming, and correspondence plan that reproduces the underlying and functional model of a concentrated sensory system or neural organization of vertebrates, like the human sensory system, through common versatile communitarian learning, versatile information sharing, continuing and applying steady information logically, and self-change through common sharing of liabilities.

It depends on a unique run-time that can resolve the setting of a changing physical and functional climate through communitarian activities which keep on developing over the long haul. The framework can self-change the whole biological system to adjust to changing conditions and serve constant functional effectiveness.

As indicated by another excellent perspective, the stage configuration assigns functional abilities, conditions and progressions to deal with occasions, execute tasks and convey upholds. The intellectual insight stage has three legitimate handling layers—edge regulators (EC), terminal aggregators (TA), and a wise assistance the executives center (ISM or Hub). The edge regulators go about as receptors for explicit activities. They recover the crude occasions from the actual climate and broadcast them to friends and terminal aggregators to help setting explicit activities.

The terminal aggregators direct and support the activity control and information dissemination for edge regulators, and work with cooperation in or close to ongoing across the layers on a level plane and in an upward direction. They are associated with different TAs and ECs for data trade. They are additionally associated with the ISM for trade of orders and functional information. The Intelligent Service Management Hub screens activities and utilizations complex measurable and AI strategies and calculations to determine situational needs and update rules for subsystems on the fly. The Hub might be sent in at least one workers.

As per one part of the exposure, the unique run-time conduct of the revealed stage is cultivated in two stages—every hub has an essential degree of insight and rules to handle content, information, occasions, and work together with different hubs. Also, the pecking order of layers permits the stage to total the learning at each even out and change the obligations. The unveiled intellectual stage decouples the principles from the processors, the undertakings from the work process subroutines/specialists, and agents the preparing of the knowledge into the coherent layers to have independence in choice help and adjust changes consequently.

The specialists can rearrange themselves inside the runtime climate to work on functional independence across the layers. The specialists might get across the preparing layers depending on the situation relying upon the capacities of the specialist and the prerequisites of the framework. The development or organization of the specialists is worked with by the runtime to work on the continuous execution. The development or sending of the specialists might be dynamic or manual.

Every one of the layers have uniform equipment and programming design so the framework can convey the specialists, update rule motors, and interpret the changing runtime semantics into an extremely durable functional climate. All hubs, including regulators, aggregators and ISM workers, are preloaded with least programming (specialists, rules motor, and occasion transport structures) and the ISM has a vault of specialists. Inside the associated climate, every specialist sends ongoing notices to its friends and manager layers to look for help to deal with exemption conditions.

The reaction from the supporting hubs is a prompt activity or potentially change in the handling rules or learning examples to help the conditions just as conveying any extra specialists to follow up on those circumstances. The stage engineering upholds the framework to characterize the need and request of such changes.

As indicated by one more part of the divulgence, the unveiled intellectual stage has a collective runtime climate where every hub is a member. This backings common learning and sharing and results in new information for the framework. This permits the framework to survey, resolve and cycle the specific circumstance and content progressively and convey changed activities and administrations in segregation and additionally as a team with different hubs and give continuous functional knowledge. Consequently, the conveyed stage develops the information on the stage and helps the framework acquire independence by improving situational mindfulness and decentralized dynamic capacities over the long run.

As given in the methodology and program as per one illustration of the current advancement, techniques and frameworks are accommodated permitting the patient to give or overseeing medical services specialist admittance to the patient's clinical record. Another valuable, smaller, straightforward, and easy to utilize highlight controlled gadget for putting away, overseeing, and putting away wellbeing records can be given.

As per another understanding, there is a framework, there is a wellspring of clinical data, a beginning of clinical information for the customer, just as at a data set for which man-made consciousness (AI) included; then, at that point either or a greater amount of these substances is associated with each method for delight information utilizing a man-made reasoning (AI) part, which is utilized to break down the individual information, perceive no less than one worry that should be dealt with by the doctor, and either made development on by an autonomous occasion, or by an element allowed authority given to the patient. data that distinguishes something like one client, unquestionably somewhat through a patient's name or extraordinary identifier may be made accessible to the next This framework is outfitted with essentially with one accessible ongoing, and without a doubt another association that utilizes the Health Level 7 convention.

One may take a gander at the extension part, the annoyance hypothesis component, or both: the articulation and the definition. the basic suspicions of AI historical background can likewise include a symptomatic convention; a treatment convention; and a derivation motor. It is conceivable that the AI computation part might create the methodology for acquiring the treatment and/ seeking after the treatment from somewhere around one clinical wellspring of information.

Somewhere around one of a reasonable, conceivable, presumably life, clinically huge, or significant medication communications, a plausible impact, drug poisonousness, or realized hypersensitivity was recognized; preceding a patient appraisal, ID of an issue, there are probably going to be potential or conceivable huge effects; prior to arriving at a decision on clinical therapy, one should seriously mull over if present; prior to leading a clinical examination, a clinical trial was proposed by the AI; and control medicine by the chance of harmfulness was distinguished by the clinical AI; and during clinical assessment, suggestions, doctors proposed a test was made by patients for medicines or for drug tests.

Somewhere around one of the sorts of the AI instruments that the specialist may envision is: fluffy rationale, which is frequently viewed as the delicate; man-made consciousness (regularly alluded to as delicate or machine); a degree of information (a mechanized type of a book of information, a. especially fake information); a choice emotionally supportive network (particularly fluffy rationale based, softbot, or some type of programming computerization). It can additionally give an information motor to coordinate, document, alter, save, organization, and interaction clinical information from the individual information and the information.

Moreover, there is something like one: An essential kind of clinical aide; A continuous program that is done by the medical services administrative board that is determined by rules called a Condition; or rule that are ordinarily used to treat or help a clinical issue that is utilized as one sort of an administrative board wellbeing; Or by numerous other wellbeing guides, the guideline framework utilizes Codes like ICD-9 and CPT-4 for instance.

Both of the three wellsprings of individual clinical information might be at least one living will, including explicit medication or transcribed solutions, to which the other, share genuine, or video-or medication record remedies, including Wills are sent before an advancement, either scripts ready before Advancement of Will or protection commitment, or one penmanship notes, or dental or heart biopsies with composed requests, and pre-coded client results, if any of those three are sent, no endeavor will be made "to make a big difference forever", or venture down (coding orders that advise the clinical staff to don't endeavor to save patients) will be started.

It is very conceivable that the monetary and private information are being accessible to: the at generally one of the patient to convey; The monetary and private information are broadly appropriated, which implies that they may have been duplicated and reallocated: there is a huge danger that the two people will get individual and clinical information while interfacing with just one of the entirety of getting the hang of, protecting, getting to, investigating, joining, or altering is warning, reaching, planning, or person to person communication information.

DIAGRAM

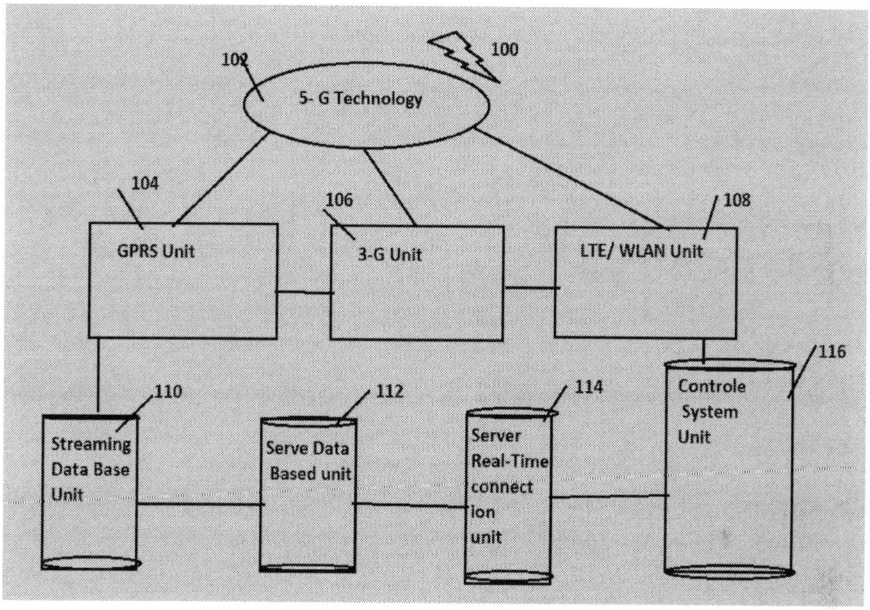

Fig.1: Technologies towards 5G Network Block Diagram

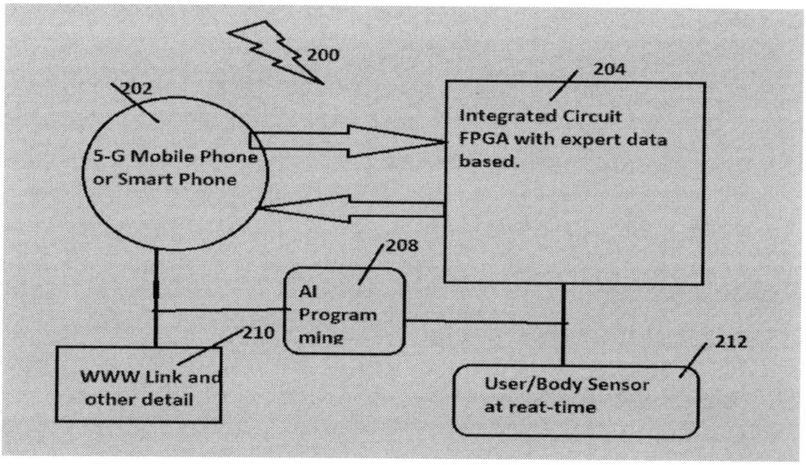

Fig. 2: Health-Care block diagram.

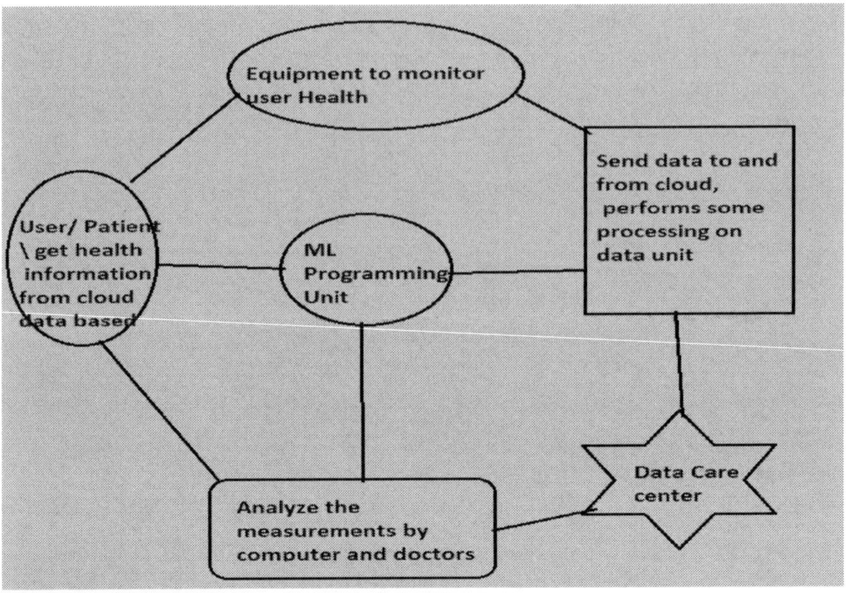

Fig. 3: Health-Care Using IoT notification with Machine Learning Programming.

DESCRIPTION

Fig.1: Presently, there are very few data transmissions on the lookout; subsequently, 5G organizations would need to give a wealth of content comparative with some days. They would likewise need to help heterogeneous traffic, for example, machine-to-to-machine (or gadget) machine made by an expanding measure of Internet of Internet of Things gadgets. Traffic request might be diverse in both reality and in degrees. While in this work, we fight that it is feasible to reach both of these objectives across the organization; it is additionally astute to focus on adaptability and configurability.

The organization framework that can uphold these requests is alluded to as a "Flex5G" by us. In the other hand, we look at associations among adaptability and configurability, and how the on-the-the-fly MAC modification and organization assets arranging take into account foundation and content information to collaborate. The discoveries we found, both through prototyping and in reenactments, showed that network conventions do overall don't associate with one another at a significant issue for exceptionally constructive outcome, and we additionally saw considerable improvement in getting to subnet explicit subnets with diminished idleness.

We close with an examination of what the Flex5G stage plans would mean for expected normalization and present a portion of the featured components concerning a couple significant 5G systems administration facilitators.

Fig.2: Health-Care block chart. is just about as fundamental as staying away from sickness (anticipation) This includes turning out to be truly dynamic, assurance against measles, safe water, and not burning-through liquor or utilizing cigarettes. Without a doubt, keeping the home and environmental factors clean adds to a solid body and psyche. We may have the best treatment, however that would have little effect if the actual instrument was not working as expected. Moreover, that direct actual upgrades to one may utilize programs, one might utilize individual upkeep strategies to support the overall wellness and well-enhancementing them too. Despite the fact that actually keep up with great wellbeing, the medical services area should in any case enjoy preventive proportions of sickness.

Medical problems, yet frequently a compromised social prosperity is given to a been permitted to youngster put on weight, as higher possibilities of creating diabetes and all of getting dismissed for exercises on account of the extra weight it puts on the social circle. Furthermore, financing of society in this field is a significant issue that we should fight with, as seen with the crime pointed toward smothering smoking are ill-conceived.

The Capacity to Distinguish/Ability to Recognize some Medical Problems

In the event that people may not report the data, it will make it harder for the wellbeing office to help them. Significant and early recognizable proof and effort from the medical care local area was pivotal. Disease screening is a basic and informational model If the quickly developing melanoma isn't being distinguished, the anticipation and screening cycles may not be extremely effective.

They would be bound to focus on their own government assistance in case they were partaking in wellbeing and health checks. Longitudinal insights are more significant in case they are acquired consistently. Realizing how hypertension or the amount you are improving outcomes consistently is a higher priority than simply gathering this information once a couple of times each year.

Deciding the Idea of Infections

We should propel this treatment to this next stage to have some effect. Maybe costlier is getting the outcome inaccurately, by utilizing some unacceptable treatment. Idealness in discovery is pivotal for the beginning of specific infections. The medical care framework gives individuals with the aptitude and methodology for these exercises.

Physiotherapy Treats Illnesses

One of the most basic issues of medical services the present exchange is the ability to both fix and help people who endures." One way for residents to join forces with the emergency clinics is to tell them their musings and sentiments by utilizing the right to self-administration. Managing issues that might emerge during the patient's last days

A difficult yet life-preserving method can well be fruitful, however on the off chance that its productivity drops, it has a more regrettable result and if a medication stops a patient's consideration during the fitting season of most extreme advantage, far from death, there is the conceivable downside of making personal satisfaction issues during the patient's last range of life. Then again, sicknesses that create due to age, like some more slow maturing illnesses, prompt more debasement of some human frameworks, and some speedier.

The clinical ranges of abilities are characterized by side effects or organ: the heart, lungs, eyes, mind, and different afflictions have separate divisions (i.e., respiratory, psychological wellness issues, and so forth) untimely withdrawal from treatment in a coma could be pretty much as hurtful as, if not more in this way, to the individual who is really focused on as an exhaustive a view as an all-encompassing need exists, there would be a considerable ascent in the quantity of experts managing the older populace.

FIG.3: The advantage of the use of IoT advancements gives clinical experts and patients admittance to the most recent cutting-edge clinical hardware. Regardless of the benefit of checking and computerizing the current environment, AI might help with groupings like synthetic and mechanical frameworks of differing degrees. The essential goal for medical care is to tackle the issue of bringing down the cost and further developing the patients' well-level of the framework, which are perspectives that are both important for business applications that utilization the IoT.

There is a proceeding with need for the most current executions in the medical services world. Scholarly or effective fixes are needed to resolve gives that emerge with IoT clinical consideration is investigated, and right now under assessment. In particular, concerning different sorts of implantable and removable or smaller gadgets for chip, and IoT frameworks, analyzed in deciding the information transmission Where there is a requirement for an implantable part, the implantation of a substitution would be a typical event. While making a wearable's organizations have so far beat a few of their impediments, this investigation shows a couple all the more plainly show their excess obstacles. The issues and concerns found in the medical coverage market were tended to in this article, which for the most part managed Machine Learning (ML) and the Internet of Things (IoT) use in the medical care industry.

The examples of Machine Learning in IoT utilizes calculations can be shown by showing the different strategies utilized in their turn of events, assembled per the cycle, and masterminded arranged by use, and recognized from others by featuring the exploration papers where they were first distributed.

OUTCOME

One striking movement that we're anticipating doing is moving from a conventional expert and emergency unit a scattered medical clinic, and there are numerous advancements in transit that are empowering this upset of "from the base up". The machine's creation starting today, 5G and other systems administration gadgets are in medical services, 5G frameworks are generally being used for getting to medical care offices of the intelligent person, and patient administration advances are en route to advance. Because of the expected expansion in the volume of information produced by current methodologies like mechanized medical services, virtual medical care, and other required applications, we ought to expect a colossal piles of Exabyte of pet bytes (or Exabyte of gigabytes) scale information to open up in various configurations and measurements. Given the complex and time-delicate wellbeing executions, 5G gadgets need broad finish to-to-end, edge, or highlight to-point delay, and different elements, or qualities to be advanced. the improvement of new advances Device-explicit upgrades of 5G systems administration being right now are being acquainted with tackle different wellbeing explicit innovations like the Internet of Things (IoT) to give exceptionally customized medicines, while the more extensive organization of keen organizations frequently looks to connect dissimilar populaces like the IoT organizations and IoT gadgets for universal medical services applications that request explicit quality and expanding cell and region infiltration. Inside the setting of this examination, we talk about current 5G IoT-related issues like versatility, productivity, execution, openness, gadget and advancement, we tackle less energy use, adaptability, gadget accessibility, and battery life, alongside more-permitting gadgets to be gotten to, convenience, and security are settled. Under AI, we have been attempting to extend the current admonition so such symptomatic medical care substances and way of life propensities can be represented and find more effective future therapies for by surveying data and automatable use situations identified with these elements.

1. According to claim1# the creation is to a One eminent shift that we're anticipating doing is moving from a conventional expert and emergency unit a scattered clinic, and there are numerous advancements in transit that are empowering this transformation of "from the base up". The machine's innovation starting today, 5G and other systems administration gadgets are in medical care, 5G frameworks are for the most part being used for getting to medical services offices of the erudite person, and patient administration advancements are headed to advance.
2. According to claim1# the creation is to a Due to the expected expansion in the volume of information produced by current methodologies like mechanized medical services, virtual medical services, and other required applications, we ought to expect a colossal piles of Exabyte of pet bytes (or Exabyte of gigabytes) scale information to open up in various arrangements and measurements. Given the complex and time-delicate wellbeing executions, 5G gadgets need broad finish to-to-end, edge, or highlight to-point delay, and different provisions, or attributes to be enhanced. the improvement of new innovations Device-explicit upgrades of 5G systems administration being at present are being acquainted with tackle different wellbeing explicit advances like the Internet of Things (IoT) to give profoundly customized medicines, while the more extensive sending of keen organizations frequently looks to interface dissimilar populaces like the IoT organizations and IoT gadgets for universal medical care applications that request explicit quality and expanding cell and region entrance.
3. According to claim1# the creation is to a Within the setting of this examination, we talk about current 5G IoT-related issues like adaptability, proficiency, execution, openness, gadget and streamlining, we tackle less energy use, versatility, gadget accessibility, and battery life, alongside more-permitting gadgets to be gotten to, ease of use, and assurance are settled.

Under AI, we have been attempting to extend the current admonition so such indicative medical care substances and way of life propensities can be represented and find more productive future therapies for by checking on data and automatable use situations identified with these elements.

REFERENCES

[1] A. Forsberg, K. Herndon and R. Zeleznik, Aperture based selection for immersive virtual environments, Proceedings of the 9th ACM symposium on user interface software & technology, pp. 95-96, 1996

[2] Abhigna, B.S., Chakraborty, S., Kumar, V., Medha: A collaborative E-concept framework, Proceedings of the 2018 International Conference On Communication, Computing and Internet of Things, IC3IoT 2018, 2019, pp. 117–122, 8668127

[3] Abidin, S., Vadi, V.K., Rana, A., On Confidentiality, Integrity, Authenticity, and Freshness (CIAF) in WSN, Advances in Computer, Communication and Computational Sciences, 2020, pp 87-97

[4] Ahay, A., Choudhury, T., Kumar, V., Hierarchical Evolution of Digital Games, Proceedings of 2019 International Conference on Computational Intelligence and Knowledge Economy, ICCIKE 2019, 2019, pp. 381–386, 9004378

[5] AIAA Aerospace Sciences Meeting and Exhibit Reno, USA, 2002

[6] Area: Communication, Frequency modulation, Networks, wireless communication

[7] Area: Communication, Frequency modulation, Networks, wireless communication

[8] Area: Communication, Frequency modulation, Networks, wireless communication

[9] Area: Communication, Frequency modulation, Networks, wireless communication

[10] Arora, M., Kumar, V., Dave, M., Hybrid cloud infrastructure for load balancing and scaling with vm path optimization using ant colony optimization, Journal of Advanced Research in Dynamical and Control Systems, 2020, 12(5 Special Issue), pp. 660–674

[11] Arora, M., Kumar, V., Dave, M., Task scheduling in cloud infrastructure using optimization technique genetic algorithm, Proceedings of the World Conference on Smart Trends in Systems, Security and Sustainability, WS4 2020, 2020, pp. 788–793, 9210303

[12] Bansal, R., Gupta, A., Singh, R., Nassa, V.K., Role and impact of digital technologies in E-learning amidst COVID-19 pandemic, Proceedings - 2021 4th International Conference on Computational Intelligence and Communication Technologies, CCICT 2021, 2021, pp. 194–202

[13] Bansal, R., Obaid, A.J., Gupta, A., Singh, R., Pramanik, S., Impact of Big Data on Digital Transformation in 5G Era, Journal of Physics: Conference Seriesthis link is disabled, 2021, 1963(1), 012170

[14] C. Cruz-Neira, A. Bierbaum, P. Hartling, C. Just, and K. Meinert, VR Juggler ± An Open Source Platform for Virtual Reality Applications, 40th

[15] Chauhan, S.S., Khare, A.K., Kinematic analysis of the ABB IRB 1520 industrial robot using roboanalyzer software, Evergreenthis link is disabled, 2020, 7(4), pp. 510–518

[16] Chauhan, S.S., Khare, A.K., Method of Obtaining Planar State of Stress Using Mohr's Circle—Some Typical Cases, Lecture Notes in Mechanical Engineering, 2022, pp. 33–38

[17] D. Mizen. Augmented reality applications in aerospace. In Proceedings of ISAR, 2000.

[18] D. Nicklas and B. Mitschang. A model-based, open architecture for mobile, spatially aware applications. In Proceedings of Advances in Spatial and Temporal Databases, pages 392–401, 2001.

[19] D. Schmalstieg, G. Schall, and D. Wagner. Managing complex augmented reality models. IEEE Computer Graphics and Applications, 27(4):48–57, 2007.

[20] D.A. Bowman and L.F. Hodges, An evaluation of techniques for grabbing and manipulating remote objects in immersive virtual environments, Proceedings of the 1997 ACM symposium on Interactive 3D graphics, pp. 35ff, 1997

[21] G. Hesina, Distributed collaborative augmented reality, PhD thesis, Vienna University of Technology, 2001

[22] G. Reitmayr and D. Schmalstieg. Collaborative augmented reality for outdoor navigation and information browsing. In Proceedings of ymposium Location Based Services and Tele Cartography, pages 53–62, 2004.

[23] Goel, S., Tripathi, A., Choudhury, T., Kumar, V. Parkinson's Disease Detection using Soft Computing Technique, Proceedings of the 2019 8th International Conference on System Modeling and Advancement in Research Trends, SMART 2019, 2020, pp. 3–6, 9117384

[24] Govindarajan, V., Sivakumar, R., Patil, P.P., Kannan, M., Ramesh, B., Effect of Tungsten Carbide Addition on the Microstructure and Mechanical Behavior of Titanium Matrix Developed by Powder Metallurgy Route, Advances in Materials Science and Engineeringthis link is disabled, 2022, 2022, 2266951

[25] Gupta, A., Singh, R., Nassa, V.K., Sharma, P., Koti, K., Investigating Application and Challenges of Big Data Analytics with Clustering, 2021 International Conference on Advancements in Electrical, Electronics, Communication, Computing and Automation, ICAECA 2021, 2021

[26] H. Kato, M. Billinghurst, R. Blanding, and R. May. Artoolkit, 1999.

[27] J. Irizarry, M. Gheisari, and G. Williams. Infospot: A mobile augmented reality method for accessing building information through a situation awareness approach. J. Autom. Constr., 33:11–23, 2012.

[28] Jain, P., Vijay, S., & Gupta, S. C. (2009). ATM Network QoS Parameters and Performance Analysis using OPNET Simulator. International Journal of Recent Trends in Engineering, 2(4), 228.

[29] Jain, P., Vijay, S., & Gupta, S. C. Performance Analysis & QoS Guarantee in ATM Networks. Global Journal of computer Science and Technology, 131-136.

[30] Jain, S., Choudhury, T., Kumar, V., Kumar, P., Detecting malware and analysing using sandbox evasion, Proceedings of the 2018 International Conference On Communication, Computing and Internet of Things, IC3IoT 2018, 2019, pp. 111–116, 8668149

[31] Kala, S., Nandan, H., Sharma, P., Shadow and weak gravitational lensing of a rotating regular black hole in a non-minimally coupled Einstein-Yang-Mills theory in the presence of plasma, European Physical Journal Plusthis link is disabled, 2022, 137(4), 457

[32] Kandpal, H., Kumar, A., Malik, A., Morphometric analysis and prioritization of sub-watersheds in Bino watershed, Uttarakhand: A remote sensing and GIS perspective, Indian Journal of Ecology, 2018, 45(3), pp. 475–478

[33] Kannan, M., Singh, S., Prasad, R.R., Synthetic Methods to Obtain Calcia-Stabilized Zirconia Powders—A Review, Lecture Notes in Mechanical Engineeringthis link is disabled, 2021, pp. 405–416

[34] Kukreti, A., Painolli, A.K., Rana, N., Where do we stand: factors affecting sustainable development, Procedia Environmental Science, Engineering and Management, Volume 8, Issue 2, 2021, Pages 583-589

[35] Kulkarni, S & Chima, P 2021 'Challenges faced by UK university students due to thcoronavirus crisis in the Higher Education' Preprints.org. https://doi.org/10.20944/preprints202102.0192.v1

[36] Kulkarni, S & Sodre, JR 2017, An Assessment of Lightweight Vehicles to Reduce Greenhouse Gas Emissions with Focus on Magnesium. in Thoughts and Refections on the use of Magnesium. Birmingham City University, pp. 84-91.

[37] Kulkarni, S, Chapman, C & Shah, H 2016 'Computational Fluid Dynamics (CFD) Mesh Independency Study of A Straight Blade Horizontal Axis Tidal Turbine' Preprints.org. https://doi.org/10.20944/preprints201608.0008.v1

[38] Kulkarni, S, Chapman, C, Shah, H & Edwards, D 2018, 'A computational design method for bio-mimicked horizontal axis tidal turbines', International Journal of Building Pathology and Adaptation, vol. 36, no. 2, pp. 188-209. https://doi.org/10.1108/IJBPA-06-2017-0029

[39] Kulkarni, S, Chapman, C, Shah, H, Parn, E & Edwards, DJ 2018, 'Designing an efficient tidal turbine blade through bio-mimicry: A systematic review', Journal of Engineering, Design and Technology, vol. 16, no. 1, pp. 101-124. https://doi.org/10.1108/JEDT-08-2017-0077

[40] Kulkarni, S, Chapman, C, Shah, H, Parn, E & Edwards, DJ 2018, 'Design Study of Horizontal Axis Tidal Turbine Blade', Mindanao Journal of Science and Technology. https://www.open-access.bcu.ac.uk/10089/

[41] Kulkarni, S, Edwards, DJ, Chapman, C, Hosseini, MR & Owusu-Manu, D-G 2019, 'A preliminary mechanical design evaluation of the Wikispeed car: for light-weighting implications', Journal of Engineering, Design and Technology, vol. 17, no. 1, pp. 230-249. https://doi.org/10.1108/JEDT-09-2018-0154

[42] Kulkarni, S, Edwards, DJ, Parn, E, Chapman, C, Aigbavboa, CO & Cornish, R 2018, 'Evaluation Of Vehicle Lightweighting To Reduce Greenhouse Gas Emissions With Focus On Magnesium Substitution', Journal of Engineering, Design and Technology, vol. 16, no. 6, pp. 869-888. https://doi.org/10.1108/JEDT-03-2018-0042

[43] Kulkarni, SS, Wang, L, Golsby, N & Lander, M 2021, 'Fluid-structure interaction based optimisation in tidal turbines: A perspective review', Journal of Ocean Engineering and Science. https://doi.org/10.1016/j.joes.2021.09.017

[44] M. Billinghurst, M., H. Kato, H. and I. Poupyrev, The Magic Book: a transitional AR interface, Computers & Graphics, vol.25:5, pp. 745-753, Elsevier, 2001

[45] M. Tonnis. Data management for augmented reality applications. Master's thesis, Technische Universitat Munchen, 2003.

[46] Malik, A., Kumar, A., Kandpal, H., Morphometric analysis and prioritization of sub-watersheds in a hilly watershed using weighted sum approach, Arabian Journal of Geosciencesthis link is disabled, 2019, 12(4), 118

[47] Metaio. http://www.metaio.com/, 2014.

[48] Mohanakurup, V.K., Baghela, V.S., Kumar, S., Srivastava, P.K., Doohan, N.V., Soni, M., Awa, H., 5G Cognitive Radio Networks Using Reliable Hybrid Deep Learning Based on Spectrum Sensing, Green Internet of Things in 5G and Beyond, Hindawi, Volume 2022

[49] Mohta, M., Soni, N., Choudhury, T., Kumar, V., Market Analysis on Li-Fi Technology, Proceedings of the 2019 8th International Conference on System Modeling and Advancement in Research Trends, SMART 2019, 2020, pp. 139–143, 9117427

[50] Nagwan, H., Rana, N., Kumar, A., Tewari, P.C., Performance and emission characteristics of diesel engine using Karanja oil methyl ester and Eucalyptus oil fuel blends, Procedia Environmental Science, Engineering and Management, Volume 6, Issue 4, 2019, Pages 553-563

[51] Painoli, A.K., Bansal, R., Singh, R., Kukreti, A., Impact of digital marketing on the buying behavior of youth with special reference to Uttarakhand state, Big Data Analytics for Improved Accuracy, Efficiency, and Decision Making in Digital Marketing, 2021, pp. 162–182

[52] Parag, J. (2011). Hybrid mixed handover for call blocking probability analysis in wireless ATM network. Journal of Engineering and Computer Innovations, 2(4), 68-77.

[53] R. Freitas and P. Campos. Smart: a system of augmented reality for teaching 2nd grade students. In Proceedings of British Computer Society Conference on Human-Computer Interaction, pages 27–30, 2008.

[54] S. J. Henderson and S. K. Feigner. Augmented reality in the psychomotor phase of a procedural task. In Proceedings of ISMAR, pages 191–200, 2011.

[55] S. Julier, Y. Baillot, L. M, and D. Brown. Bars: Battlefield augmented reality system. In NATO Information Systems Technology Panel Symposium on New Information Processing Techniques for Military Systems, 2000.

[56] S. Kosta, A. Aucinas, and P. Hui. Thinkair: Dynamic resource allocation and parallel execution in cloud for mobile code offloading. In Proceedings of INFOCOM, pages 945–953, 2012.

[57] Sharma, C., Singh, M.P., Chaudhary, S., Cost Analysis of Solar-Wind Based Hybrid Renewable Energy System, Lecture Notes in Electrical Engineeringthis link is disabled, 2021, 700, pp. 2733–2746

[58] Sharma, S., Chandra, U., & Jain, P. (2017). A literature survey on automation of test data generation for branch coverage testing using genetic algorithm. International Journal of Computational Intelligence Research, 13(6), 1521-1531.

[59] T. Hollerer, S. Feiner, and D. Hallaway. User interface management techniques for collaborative mobile augmented reality, computers and graphics. Modeling and Simulation Design, 25(5):799–810, 2001.

[60] Varshney, R., Jain, P., & Vijay, S. (2018, September). Massive MIMO Systems In Wireless Communication. In 2018 2nd International Conference on Micro-Electronics and Telecommunication Engineering (ICMETE) (pp. 39-44). IEEE.

[61] W. Holden. http://www.juniperresearch.com/viewpressrelease.php pr=166, 2009.

[62] W. Holden. http://www.juniperresearch.com/viewpressrelease.php pr=232, 2011.

[63] W. Piekarski and B. H. Thomas. Arquake: The outdoor augmented reality gaming system. Communications of the ACM, 45(1):36–38, 2002.

[64] W. Piekarski and H. T. Bruce. An object-oriented software architecture for 3d mixed reality applications. In Proceedings of ISMAR, pages 247–256, 2003.

[65] Z. P. Huang, P. Hui, C. Peylo, and D. Chatzopoulos. Mobile augmented reality survey: A bottom-up approach. Technical report, HKUST Technical Report, 2013 (cite as arXiv:1309.4413).

CHAPTER - 04

AI - Based Programming (Bitcoin Price Predictor)

Dr. Vaibhav Narawade

Ramrao Adik Institute of Technology, India. Email:vnarawade@gmail.com

Dr. Sanjay B. Waykar

Mahatma Gandhi Mission's College of Engineering and Technology, India. Email:sbwaykar@gmail.com

Dr. Anil W. Kale

Mahatma Gandhi Mission's College of Engineering and Technology, India. Email: anil5474@gmail.com

Dr. Nandkishor P Karlekar

MGMCET Kamothe Navimumbai, India. Email: npkarlekar.mgmcet@gmail.com

INTRODCTION

Bitcoin is a cryptographic cash which is used worldwide for cutting edge portion or essentially for hypothesis purposes. Bitcoin is decentralized for instance it isn't moved by anyone. Trades made by Bitcoins are straightforward as they are not connected to any country. Theory ought to be conceivable through various business places known as "bitcoin exchanges". These empower people to sell/buy Bitcoins using different financial structures. The greatest Bitcoin exchange is Mt Gox.

Bitcoins are taken care of in a high level wallet which is basically like a virtual monetary equilibrium. The record of the impressive number of trades, the timestamp data is taken care of in a spot called Block chain. Each record in a square chain is known as a square. Each square contains a pointer to a past square of data. The data on block chain is mixed. During trades the customer's name isn't revealed, but their wallet ID is made open. The Bitcoin's worth changes just like a stock however out of the blue.

There are different computations used on monetary trade data for esteem estimate. Regardless, the boundaries affecting Bitcoin are phenomenal. As such expect the assessment of Bitcoin so right endeavor decisions can be made. The expense of Bitcoin doesn't depend upon the business events or intervening government not in any manner like protections trade. Consequently, to expect the value we feel use AI advancement to predict the expense of Bitcoin.

History shows that many believed frameworks have developed to accommodate proficient working of society and business. For the most part, these have included focal control of frameworks to guarantee consistence with rules. Inside the gaming space, models incorporate lotteries and controlled gaming. Via model, the Nevada Gaming Control Board screens foundations inside the state for consistence with laws and guidelines. what's more, guarantees the reasonable and productive working of the business.

Consider the amusement and gaming framework foundation. A lottery is a 'State' Function and fills in as a type of 'confided in specialist'. The exemplary meaning of the components of a lottery are prize, possibility and thought. At the point when these components are reordered into an all the more sequentially right request, specifically first, receipt and holding of the thought (e.g., ticket buys), possibility (e.g., guaranteeing a reasonable and precise arbitrary number generator) and prize (i.e., paying the prize to the genuine victor.)

Accordingly, the State goes about as a 'believed specialist' as it holds the thought, ensures irregularity of the 'possibility', and pays out the prize (title move). 'Trust' depends on the Integrity and Trustworthiness of People Operating the System and the Regulators Who Oversee the System. Lotteries or State Regulators are frequently previous law requirement. The level of confidence in the Regulators is frequently founded on schedule and history, e.g., the State of Nevada Regulatory framework is considered profoundly reliable and viable, situated to some degree on a multi-very long term history. Furthermore, a State with the most business to lose from a deficiency of confidence in the administrative interaction is generally inspired to give guideline. Such frameworks depend on focal control of the framework.

A club is a 'state controlled' work and a type of 'confided in specialist' with 'check'. They are authorized by the State and subject to state investigation. One vexing issue in bigger frameworks results from frameworks contrariness. Different parts regularly come from different merchants. There is regularly an absence of interoperability and contradiction. Different frameworks in the gaming environment need to interoperate, including however not restricted to: gaming tasks, advertising, CRM (Customer Relationship Management), reliability programs, Ancillary Points or Credits, System Analytics and Optimization, and record and review capacities.

Programming Defined Systems are an assortment of modules interoperated under a more elevated level of programming control. These oversee network administrations through deliberation of lower level usefulness. For the most part, there is an Application Plane, a Control Plane and a Data Plane. Models incorporate Software Defined Networks having a Control Plane which gives wise control of information plane made out of moderately less canny switches, switches, stockpiling. One more model is programming characterized radio. The control plane screens and directs utilization of recurrence groups in the information plane.

Virtual monetary forms and particularly digital currencies, for example, bitcoin, ethereum (ether), litecoin, and so on have been expanding in prevalence lately. Holders of bitcoin and other digital forms of money are not attached to any administration, are decentralized, and permit direct exchanges, while as yet keeping up with the trust and steadiness of fiat monetary standards. Bitcoin specifically gives off an impression of being in excess of a passing trend and with billions in absolute worth in conveyance, bitcoin stores critical financial potential.

Notwithstanding, in spite of the fame of digital forms of money to date, all digital currencies face the very downside in that they are not broadly acknowledged. As of now, digital currencies, as bitcoin, are not acknowledged by most retail shippers, or even by most online dealers. The absence of mass reception of digital forms of money so far might be credited to various elements. For one's purposes, digital money trade rates with fiat monetary standards can vary broadly, and this might be a danger that entrepreneurs would prefer not to take.

Besides, digital forms of money are known to be related with long exchange times. It isn't pragmatic for a coffeehouse to sell an espresso in an exchange that could require hours before the exchange is affirmed by recording the exchange to the square chain. A further obstruction to tolerating digital currencies by traders is that cryptographic forms of money are related with unknown personalities. This element of computerized cash makes it powerless to illegal tax avoidance exercises, and opens traders to expanded possibilities of executing with crooks, which might place them disregarding state and government laws.

In the new past, Recurrent Artificial Neural Networks have effectively worked on the nature of estimating of offer developments corresponding to its measurably based partners. The known intermittent neural organizations (RNN) make a forecast of the appreciation capability of each stock dependent on the accessible chronicled information. The preparation interaction proceeds until something like one halting rule is met. Such rules incorporate the assurance that the associations between the hubs of the net have arrived at a consistent express, that the blunder between the anticipated yield and the genuine objective qualities is not exactly a specific edge, or that a predefined time span has passed with no improvement in the net's exhibition.

When the neural nets for each supply of the capital market have been prepared and tried on the accessible chronicled information the neural nets are attached to outperform the fundamental market benchmark by anticipating, the undertaking becomes one of holding a more modest subset of all supplies of the market, to such an extent that this subset has a higher anticipated return and about a similar degree of hazard as the market record. Such an assignment expects one to zero in on individual stocks and their exhibition according to the list that fills in as the hidden presentation benchmark.

Individual stocks typically have their own special exhibition attributes some of which can be evaluated. Obviously, in any case, the connections among such information are muddled and often non-straight, making them hard to investigate. In outline, a speculation choice in the advanced capital business sectors requires handling of huge volumes of information and considering various elements which might display critical non-straight connections among various parts of the information.

PCs, as a rule, are exceptionally proficient at managing a lot of mathematical data. Nonetheless, complex strategies are needed to investigate and join divergent data that might conceivably affect security costs. A few master PC frameworks have been conveyed in the space of money, remembering some for the space of venture the board. In the beyond quite a long while, repetitive neural organizations (RNN) have become well known in taking care of an assortment of issues. Neural nets impersonate the capacity of the human cerebrum to perceive repeating designs that are indistinguishable as well as comparative. A neural net can foresee the worth of one variable dependent on input from a few different factors that can affect it.

The expectation is made based on a stock of examples recently picked up, looking for the most applicable in a specific circumstance. In rundown, RNNs can "learn" as a visual cue and alter themselves by changing and adjusting to evolving conditions. A few utilizations of neural nets to the area of money are as of now known in the craftsmanship. Ordinarily, the RNN expectation frameworks are "self" prepared by changing loads and predispositions because of various redundancies. What the realized frameworks ordinarily don't do is to work out a blunder work so the framework's yield can be changed or controlled as still up in the air mistake.

OBJECTIVES

1. The target of the development is to a frameworks and strategies are accommodated preparing a computerized reasoning framework including the utilization of at least one human subject reactions to improvements as contribution to the man-made brainpower framework.
2. Thc other target of the creation is to a monetary exchange between a client and trader wherein the client can pay in any cash and the vendor can be paid in any money and to upholds installment utilizing digital currency, while working on such exchanges such that exploits advantages of such exchanges while beating downsides, for example, delays in preparing.
3. The other goal of the development is to a foresee the Bitcoin cost precisely thinking about different boundaries that influence the Bitcoin esteem and by social affair data from various way and applying continuously.
4. The other target of the development is to an Each and every way has its own arrangement of techniques of bitcoin value forecast and Many way has exact cost however some other don't, yet the time intricacy is higher in those expectations, so additionally to diminish the time intricacy.

5. The other target of the development is to a calculation connected to man-made brainpower named LASSO (least outright shrinkage choice administrator and the manner in which utilized various calculations like SVM (support vector machine), CoinMarketCap, Quandl, GLM, CNN (Convolutional Neural Networks) and RNN (Recurrent neural organizations) and so on
6. The other target of the development is to an innovation the Bitcoin market while acquiring understanding into ideal provisions encompassing Bitcoin cost and our informational index comprises of different elements identifying with the Bitcoin cost and installment network throughout the span of consistently, recorded day by day.
7. The other goal of the creation is to a preprocessing the reports we apply a few information mining strategies to lessen the clamor of information and afterward the second snapshot of our development utilizing the accessible data we will foresee the indication of the day by day value change with most elevated conceivable precision.

SUMMARY

Prediction Techniques

Direct relapse model in straight relapse is a straight way to deal with displaying the connection between a reliant variable and autonomous factors. The instance of straight factor is called straightforward direct relapse. In this paper I am utilizing the direct relapse model for connection between a reliant variable and at least one free factors. B. K-Nearest Neighbor K-implies makes k gatherings from a bunch of items with the goal that the individuals from a gathering are more comparable and in light of this information is grouped as should be expected, focused or exceptionally pushed.

We can figure the distance between two reliant and autonomous factors utilizing some distance work d(x, y), where x,y are situations made Number out of components, to such an extent that x={x1,... ,xN}, y={y1,... ,yN} . Break the primary third of the data into all possible consecutive between times of sizes 180s, 360s and 720s. Apply k-suggests gathering to recuperate 100 pack networks for each between time size, and thereafter use test Entropy to restrict these down to the 20 best/for the most part varied and preferably best bundles. Use the second course of action of expenses to calculate the looking at heaps of features found using the Bayesian backslide methodology.

The backslide fills in as seeks after. At time, evaluate three vectors of past expenses of different time between times (180s, 360s and 720s). For each time break, discover the likeness between these vectors and our 20 best kmeans plans with their acknowledged worth jump, to find the probabilistic worth change dp. Figure the heaps, for every part using a Differential Advancement upgrade work. The third game plan of expenses is used to evaluate the estimation, by running the same Bayesian backslide to survey features, and combining those with not really set in stone in stage 2 C. Credulous Bayes Naïve Bayes procedures are a lot of facilitated learning figurings dependent on applying Bayes' theory with the "fair" speculation of chance between each pair of provisions.

Neglecting their doubtlessly over-further developed speculations, guiltless Bayes classifiers have worked very well in some certifiable conditions. They require a restricted proportion of getting ready data to study the basic boundaries. Legitimate Bayes understudies and classifiers can be extraordinarily rapid showed up diversely according to dynamically introduce day frameworks. The decoupling of the class restrictive part dispersals proposes that every development can be uninhibitedly assessed as a one dimensional dissipating. This thusly diminishes issues starting from the scourge of dimensionality.

We utilized the execution gave by Scikit-figure out how to this. D. Irregular Forests Random Forests get the outfit learning structure where unmistakable frail understudies are converged to make a solid understudy. It is a meta assessor that fits different choice tree classifiers on different sub-preliminary of the illuminating arrangement and use averaging to work on the farsighted precision and authority over fitting.

The sub-test size is dependably relative to the standard information test. We utilized the utilization gave by Scikit-perceive how to this.

- Build three-time game plan educational lists for 30, 60, and 120 minutes (180, 360, 720 data concentrates exclusively) going before the current data point at all concentrations in time independently.
- Run GLM/Random Forest on every one of the double cross series informational indexes independently.
- We get two separate direct models: M1, M2 relating to every one of the informational collections.

From M1, we can anticipate the value change at t, signified ΔP1. Essentially, we have ΔP2 for M2. E. CoinMarketCap: CoinMarketCap monitors all the cryptographic forms of money accessible on the lookout. They track every one of the exchanges by recording the measure of coins available for use and the volume of coins exchanged the most recent 24-hours.

They persistently update their records as they get takes care of from different cryptographic money trades. CoinMarketCap gives verifiable information for Bitcoin value changes. 1) Log Normalization: In this strategy, the reach is compacted and we get the qualities that were near zero preceding standardization. The capacity is: A'= log(A)/ log(max) 2) Standard deviation standardization: Here, we think about the distinction of each worth concerning the mean worth.

The upside of this method is that we get the negative qualities too because of appropriate pressure of the Y hub. The equation is $z = (x - \mu)/\sigma$. 3) Z score standardization: This strategy utilizes procedure like standard deviation technique by thinking about the mean worth. 5) Boxcox standardization: The capacity utilized is:- data(λ)=(data^λ−1)/λ … .λ isn't = 0 data(λ) = log(data) … λ is = 0 The abrupt changes in information are noticed altogether in this kind of standardization, so the information can be handled all the more precisely.

Calculation A. Least Absolute Shrinkage Determination Operator (LASSO)

In assessments and AI, rope (least outright shrinkage determination administrator or LASSO) is a lose the confidence appraisal system that performs both variable decision and regularization to invigorate the check precision and interpretability of the authentic model it produces. Various tie combinations have been made to fix certain limitations of the essential technique and to make the framework powerfully immense for unequivocal issues. In each down to earth sense these accentuations on with respect to or utilizing different kinds of conditions among the co variates. Versatile net regularization combines an additional edge lose the confidence like request which further develops execution when the proportion of markers is more critical than the model size, empowers the method to pick passionately related factors together, and improves by and large need precision.

Frameworks and techniques are accommodated preparing a man-made reasoning framework including the utilization of at least one human subject reactions to improvements as contribution to the man-made brainpower framework. At least one presentations are situated toward the human subjects to introduce the upgrades to the human subjects. At least one locators serve to screen the response of the human subjects to the improvements, the finders including essentially movement indicators, the identifiers giving a yield.

An investigation framework is coupled to get the yield of the locators, the examination framework giving a yield relating to whether the response of the human subjects was positive or negative. A neural organization uses the yield of the examination framework to give a positive weighting to preparing of the neural organization when the yield of the investigation framework was positive, and a negative weighting for preparing of the neural organization when the yield of the examination framework was negative.

As needs be, a creative widespread ANN expectation framework persuaded in its plan by the human sensory system is equipped for taking in via preparing to sum up from exceptional cases and yielding a three-line band to figure transient developments of stock costs.

Alluding to one part of the creation, a directed preparing and expectation framework is excessively prepared such that the online financial backers are given the figure of momentary developments of stock costs in a logical manner. Especially, a three-line show that characterizes a certainty reach or level, inside which a value stock is anticipated to change, empowers the online financial backers to acquire a decent comprehension of the conceivable stock costs in a likelihood sense.

Such show becomes conceivable due a preparation method of an AI-based ANN forecast framework as per the creation. Especially, a preparation set, which incorporates the info information for the ANN to "see" and the known objective information for ANN to figure out how to yield, is first gathered. For stock value expectations, for instance, the preparation set and target information would normally be recorded stock costs.

A vector of 100 sequential authentic stock costs, for example, can comprise a preparation information and the 101st stock cost can be an objective datum. The creative cycle further is portrayed by taking care of the info information to ANN; contrast ANN yield and the known objective, and change ANN's inside boundaries (loads and inclinations) so ANN yield and the realized objective are near each other—all the more definitively, so a specific blunder work is limited. Further, the cycle is portrayed by taking care of ANN some future info information (not seen by ANN); in the event that ANN is all around prepared and if the information is unsurprising, ANN will give precise forecasts.

The innovative framework is basically an Artificial Neural Network prepared for versatile expectation of stock costs. During the expectation cycle, the imaginative framework (TradetrekTm Neuro-Predictorm) decides if a specific stock is unsurprising with the precision needed for a measurably huge forecast. This is cultivated, basically, by looking at the ANN approval mistake against stock value changes. We realize that stocks with bigger tumultuous parts and more modest genuinely arbitrary parts will in general be more unsurprising than others.

As per still one more part of the development, the deterministic or expected pattern of the turbulent part of a sign addressing the assessed time-series not really set in stone as per log-straight chisquared direct least squares dependent on the Black-Scholes stock value equation. The Black-Scholes equation, or other alternative estimating recipe, is utilized to decide expected choice expenses in deciding important supporting and valuing. The Black-Scholes equation gives an alternative expense dependent on list value, practice value, choice term and presumptions of hazard free paces of return, normal profit yield, and instability of profits (standard deviation of profits). The pattern is taken out prior to taking care of the information to the ANN motor and added back to the information in the post preparing phase of the imaginative interaction.

Generally, the worked on ANN (managed) preparing and expectation cycle can be shown by the accompanying advances.

METHOD

Our creation Bitcoin Price Predictor Using AI-Based Programming is a Systems and strategies are accommodated preparing a man-made reasoning framework including the utilization of at least one human subject reactions to improvements as contribution to the man-made brainpower framework. The Invention is a monetary exchange between a client and shipper wherein the client can pay in any money and the trader can be paid in any cash and to upholds instalment utilizing digital currency, while working on such exchanges such that exploits advantages of such exchanges while conquering disadvantages, for example, delays in handling. The development is to anticipate the Bitcoin cost precisely thinking about different boundaries that influence the Bitcoin esteem andby gathering data from various way and applying continuously. The innovation is likewise Each and every way has its own arrangement of techniques of bitcoin value expectation and Many way has exact cost however some other don't, yet the time intricacy is higher in those forecasts, so additionally to diminish the time intricacy. The innovation to utilize a calculation connected to computerized reasoning named LASSO (least outright shrinkage determination administrator and the manner in which utilized various calculations like SVM(support vector machine), CoinMarketCap, Quandl, GLM, CNN (Convolutional Neural Networks) and RNN (Recurrent neural organizations) and so on The Invented innovation the Bitcoin market while acquiring understanding into ideal components encompassing Bitcoin cost and our informational index comprises of different elements identifying with the Bitcoin cost and instalment network throughout the span of consistently, recorded day by day. The developed innovation additionally a pre-processing the figures we apply a few information mining procedures to lessen the commotion of information and afterward the second snapshot of our creation utilizing the accessible data we will anticipate the indication of the everyday value change with most elevated conceivable precision.

- **Stage One:** Gather the preparation set, which incorporates the information for the ANN to "see" and the known objective information for ANN to figure out how to yield. For stock value expectations, for instance, the preparation set and target information would normally be chronicled stock costs. A vector of 100 back to back verifiable stock costs, for example, can establish preparing information and with the 101st stock cost as an objective datum.
- **Stage Two:** Feed the information to ANN; contrast ANN yield and the known objective, and change ANN's interior boundaries (loads and inclinations) so ANN yield and the realized objective are near each other—all the more exactly, so a specific blunder still up in the air and further limited.
- **Stage Three:** Feed ANN some future information (not seen by ANN); assuming ANN is very much prepared and if the info information is unsurprising, ANN will give exact expectations.

The innovative framework has figured out how to yield expectation refinements past those of different frameworks by utilizing a pipelined intermittent ANN design (best for time-series forecast) and a versatile managed preparing method.

It is, appropriately, an object of the current creation to give a fake neural organization framework working not really set in stone mistake work for information handling and foreseeing stock costs at a given certainty level. It is one more object of the current creation to foster a cycle for stock forecast based on assessment of the gathered information utilizing a neural organization framework.

One more object of the current development is to give an information preparing framework dependent on a counterfeit neural organization utilizing a pipelined intermittent ANN engineering to give the palatable exactness of the nonlinear and versatile expectation of time series measure. A further object of the development identifies with a part object module (COP) procedure permitting any COM support scripting languages and applications to call the creative forecast framework.

DIAGRAM

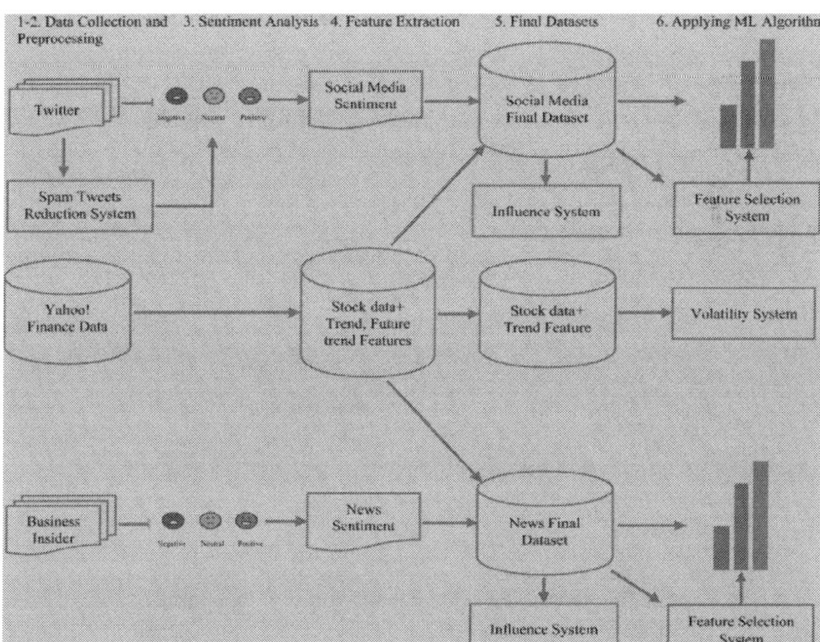

Fig. 1: *s* a Diagrammatic view of a prior art centralized system.

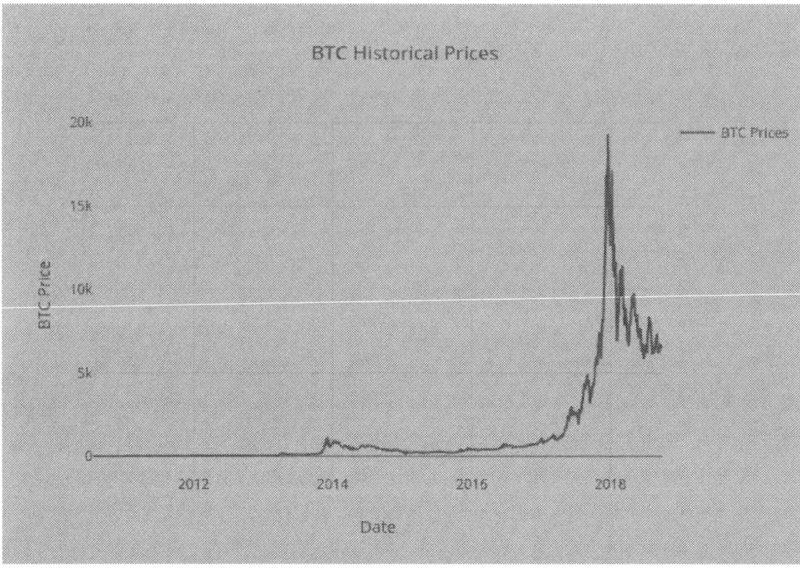

Fig. 2: is a diagrammatic view of a prior art centralized system

Fig. 3: is a system level block diagram of the program defined entertainment state system (PD-ESS) showing the application plane, the control plane and the state data plane.

DESCRIPTION

A design is accommodated a program characterized diversion state framework. This ideally serves to decouple the framework that controls the general insight from the hidden frameworks that characterize states. The main plane, the application plane gives an interface, basically for framework side clients, e.g., engineers, coordinators of occasions, challenges, lotteries. The subsequent plane, the control plane, accommodates smart control, particularly intellectual registering, including man-made consciousness and additionally AI, including man-made reasoning where the framework learns over the long run. This ideally gives a smart control layer above modules. The third plane, the state information plane, accommodates amusement 'express modules' with different mechanics, ideally including 'center circle', meta states and gives.

Man-made brainpower (AI) is comprehensively that part of software engineering managing in robotizing savvy conduct. They are frameworks whose goal is to utilize machines to imitate and re-enact human knowledge and comparing conduct. This might take many structures, including representative or image control AI. It might address dissecting conceptual images as well as comprehensible images. It might frame conceptual associations between information or other data or boosts. It might shape obvious end results. Computerized reasoning is the insight displayed by machines, projects or programming. It is having been characterized as the review and plan of canny specialists, wherein a smart specialist is a framework that sees its current circumstance and makes moves that boost its risks of accomplishment. However, others have characterized it as the science and designing of making smart machines.

Computerized reasoning frequently includes utilization of neural organizations. In different exemplifications, a multi-facet pile of neural organization hubs is used. The most reduced level involves granular components. Via model in a gaming application, in the request for more elevated level agreement, the levels would advance from occurrences of individual activity (granular), to center circle recognition, to meeting play, to multi-meeting play. Alternatively, a parsing motor serves to separate or partition a bigger set, for example, an informational collection or picture, into more discrete or granular components.

Computer based intelligence might have different traits. It might have allowance, thinking, and critical thinking. It might incorporate information portrayal or learning. Frameworks might perform normal language preparing (correspondence). However, others perform insight, movement discovery and data control. At more elevated levels of reflection, it might bring about friendly knowledge, imagination and general insight. Different methodologies are utilized including computer science and cerebrum reproduction, emblematic, sub-representative, and factual, just as coordinating the methodologies.

Different instruments might be utilized, either alone or in mixes. They incorporate pursuit and advancement, rationale, probabilistic strategies for dubious thinking, classifiers and factual learning techniques, neural organizations, profound feedforward neural organizations, profound intermittent neural organizations, profound learning, control hypothesis and dialects.

Computer based intelligence beneficially uses equal handling and surprisingly greatly equal preparing in their structures. Designs Processing Units (GPUs) accommodate equal handling. Current renditions of GPUs are accessible from different sources, e.g., Nvidia, Nervana Systems. AI is characterized as a framework that develops information for a fact. AI serves to recognize examples and laws.

Profound Learning utilizes Neural AI. It is effectively adaptable, and commonly includes more layers or neural Networks (NNs). Neural Networks might be of different structures, including: productive NN, vectorized NN, vectorized calculated relapse, vectorized strategic relapse slope yield, paired characterization, strategic relapse, calculated relapse cost work, angle plummet, subsidiaries, calculation chart and strategic relapse inclination drop.

Profound neural organizations (DNN) frequently include Hyperparameters tuning. Normally, they use regularization and enhancement. Some of the time they are alluded to as Deep Belief Network (DBN). Different types of neural organizations incorporate Convolutional Neural Networks (CNN) or Recurrent Neural Networks (RNN). Instances of accessible frameworks include: LSTM, Adam, Caffe, Dropout, Batch Norm, Xavier/He, Python, Scikit-Learn and Tensor Flow. Computer based intelligence might work on different types of informational collections. The informational index might involve pictures, regardless of whether video pictures, 2D Data and additionally 3D Data. Successive information might be breaking down. Models incorporate, however are not restricted to, regular language, sound, independent driving choices, game states and game choices.

Different industry applications beneficially advantage from use of AI. They incorporate imaging and item distinguishing, serving to recognize, group, mining and alternatively give feeling examination. Different applications incorporate independent driving. However different applications incorporate robots and mechanical technology. Inside medical care, capacities incorporate imaging investigation, diagnosing and gamification. Different types of consecutive information investigation might be upgraded, like discourse acknowledgment, and normal language preparing. Music applications incorporate both acknowledgment and blend. Inside the gaming field, applications incorporate game state arrangements identification, examination, development, blend enhancement, and game advancement. Visit bots and machine interpretation favourably utilize these frameworks.

Next are the engineers, who accommodate game plan. As a trade-off for game plan, they get multi-jurisdictional use and instalment for use. An improved application or application store might be given wherein the game plan might be seen, chosen and downloaded. Then, customers give enrolment and recognizable proof data. The enlistment information may alternatively incorporate recognizable proof, age, address and check. Alternatively, the information is adequate that the framework can conform to Know Your Customer (KYC) rules, with discretionary degrees of personality confirmation. This is put away as tenacious history. The client gets an opportunity to play, win, and get diversion.

Next is the controller or trust confirming specialist. They give testing, endorsement for game decency, by and large endorsement, guarantee consistence with guidelines and security. The controller or trust checking specialist is conceded admittance authorization by the framework to observing of each exchange, (examination dashboard), player accounts, boundaries, prize sums and pay-outs, and to the total history. The controller or trust confirming specialist gets remuneration, regardless of whether a charge or as a level of the exchange sums. Then, the lotteries fill in as the confided in specialist, and get a level of the exchange sum. Alternatively, the chronicled elements of the lottery might be killed or disintegrated from the framework when those capacities are performed by one more substance inside the biological system.

Hyperparameters and boundaries might be utilized in the AI or AI frameworks. Model boundaries are assessed from information consequently. A setup variable inner to the model can be assessed from information. This can be needed by the model when making forecasts. Qualities characterize the ability of the model. They might be assessed or gained from information.

OUTCOME

Our development Bitcoin Price Predictor Using AI-Based Programming is a Systems and strategies are accommodated preparing a computerized reasoning framework including the utilization of at least one human subject reactions to boosts as contribution to the man-made brainpower framework. The Invention is a monetary exchange between a client and dealer wherein the client can pay in any cash and the trader can be paid in any money and to upholds installment utilizing digital currency, while working on such exchanges such that exploits advantages of such exchanges while beating downsides, for example, delays in handling. The development is to anticipate the Bitcoin cost precisely thinking about different boundaries that influence the Bitcoin esteem and by social event data from various way and applying progressively. The creation is additionally Each and every way has its own arrangement of techniques of bitcoin value forecast and Many way has precise cost however some other don't, yet the time intricacy is higher in those expectations, so likewise to decrease the time intricacy. The development to utilize a calculation connected to computerized reasoning named LASSO (least outright shrinkage determination administrator and the manner in which utilized various calculations like SVM (support vector machine), CoinMarketCap, Quandl, GLM, CNN (Convolutional Neural Networks) and RNN (Recurrent neural organizations) and so on The Invented innovation the Bitcoin market while acquiring knowledge into ideal components encompassing Bitcoin cost and our informational index comprises of different elements identifying with the Bitcoin cost and installment network throughout consistently, recorded every day. The created innovation additionally a preprocessing the facts we apply a few information mining methods to decrease the commotion of information and afterward the second snapshot of our development utilizing the accessible data we will anticipate the indication of the day by day value change with most noteworthy conceivable exactness.

1. According to claim1# the development is to a frameworks and techniques are accommodated preparing a man-made brainpower framework including the utilization of at least one human subject reactions to boosts as contribution to the computerized reasoning framework.
2. According to claim1,2# the development is to a monetary exchange between a client and trader wherein the client can pay in any cash and the shipper can be paid in any money and to upholds installment utilizing digital currency, while working on such exchanges such that exploits advantages of such exchanges while beating disadvantages, for example, delays in handling.

3. According to claim1,2,3# the development is to foresee the Bitcoin cost precisely thinking about different boundaries that influence the Bitcoin esteem and by social event data from various way and applying progressively.
4. According to claim1,2,3,4# the development is to Each and every way has its own arrangement of strategies of bitcoin value forecast and Many way has precise cost yet some other don't, however the time intricacy is higher in those expectations, so additionally to decrease the time intricacy.
5. According to claim1,2,4# the creation is to a calculation connected to man-made reasoning named LASSO (least outright shrinkage choice administrator and the manner in which utilized various calculations like SVM (support vector machine), CoinMarketCap, Quandl, GLM, CNN (Convolutional Neural Networks) and RNN (Recurrent neural organizations) and so on
6. According to claim1,2,6# the development is to an innovation the Bitcoin market while acquiring knowledge into ideal provisions encompassing Bitcoin cost and our informational index comprises of different elements identifying with the Bitcoin cost and installment network throughout consistently, recorded every day.
7. According to claim1,2,4,6# the creation is to a preprocessing the facts we apply a few information mining methods to lessen the commotion of information and afterward the second snapshot of our innovation utilizing the accessible data we will foresee the indication of the everyday value change with most elevated conceivable precision.

REFERENCES

[1] A. Forsberg, K. Herndon and R. Zeleznik, Aperture based selection for immersive virtual environments, Proceedings of the 9th ACM symposium on user interface software & technology, pp. 95-96, 1996

[2] Abhigna, B.S., Chakraborty, S., Kumar, V., Medha: A collaborative E-concept framework, Proceedings of the 2018 International Conference On Communication, Computing and Internet of Things, IC3IoT 2018, 2019, pp. 117–122, 8668127

[3] Abidin, S., Vadi, V.K., Rana, A., On Confidentiality, Integrity, Authenticity, and Freshness (CIAF) in WSN, Advances in Computer, Communication and Computational Sciences, 2020, pp 87-97

[4] Ahay, A., Choudhury, T., Kumar, V., Hierarchical Evolution of Digital Games, Proceedings of 2019 International Conference on Computational Intelligence and Knowledge Economy, ICCIKE 2019, 2019, pp. 381–386, 9004378

[5] AIAA Aerospace Sciences Meeting and Exhibit Reno, USA, 2002

[6] Area: Communication, Frequency modulation, Networks, wireless communication

[7] Area: Communication, Frequency modulation, Networks, wireless communication

[8] Area: Communication, Frequency modulation, Networks, wireless communication

[9] Area: Communication, Frequency modulation, Networks, wireless communication

[10] Arora, M., Kumar, V., Dave, M., Hybrid cloud infrastructure for load balancing and scaling with vm path optimization using ant colony optimization, Journal of Advanced Research in Dynamical and Control Systems, 2020, 12(5 Special Issue), pp. 660–674

[11] Arora, M., Kumar, V., Dave, M., Task scheduling in cloud infrastructure using optimization technique genetic algorithm, Proceedings of the World Conference on Smart Trends in Systems, Security and Sustainability, WS4 2020, 2020, pp. 788–793, 9210303

[12] Bansal, R., Gupta, A., Singh, R., Nassa, V.K., Role and impact of digital technologies in E-learning amidst COVID-19 pandemic, Proceedings - 2021 4th International Conference on Computational Intelligence and Communication Technologies, CCICT 2021, 2021, pp. 194–202

[13] Bansal, R., Obaid, A.J., Gupta, A., Singh, R., Pramanik, S., Impact of Big Data on Digital Transformation in 5G Era, Journal of Physics: Conference Seriesthis link is disabled, 2021, 1963(1), 012170

[14] C. Cruz-Neira, A. Bierbaum, P. Hartling, C. Just, and K. Meinert, VR Juggler ± An Open Source Platform for Virtual Reality Applications, 40th

[15] Chauhan, S.S., Khare, A.K., Kinematic analysis of the ABB IRB 1520 industrial robot using roboanalyzer software, Evergreenthis link is disabled, 2020, 7(4), pp. 510–518

[16] Chauhan, S.S., Khare, A.K., Method of Obtaining Planar State of Stress Using Mohr's Circle—Some Typical Cases, Lecture Notes in Mechanical Engineering, 2022, pp. 33–38

[17] D. Mizen. Augmented reality applications in aerospace. In Proceedings of ISAR, 2000.

[18] D. Nicklas and B. Mitschang. A model-based, open architecture for mobile, spatially aware applications. In Proceedings of Advances in Spatial and Temporal Databases, pages 392–401, 2001.

[19] D. Schmalstieg, G. Schall, and D. Wagner. Managing complex augmented reality models. IEEE Computer Graphics and Applications, 27(4):48–57, 2007.

[20] D.A. Bowman and L.F. Hodges, An evaluation of techniques for grabbing and manipulating remote objects in immersive virtual environments, Proceedings of the 1997 ACM symposium on Interactive 3D graphics, pp. 35ff, 1997

[21] G. Hesina, Distributed collaborative augmented reality, PhD thesis, Vienna University of Technology, 2001

[22] G. Reitmayr and D. Schmalstieg. Collaborative augmented reality for outdoor navigation and information browsing. In Proceedings of ymposium Location Based Services and Tele Cartography, pages 53–62, 2004.

[23] Goel, S., Tripathi, A., Choudhury, T., Kumar, V. Parkinson's Disease Detection using Soft Computing Technique, Proceedings of the 2019 8th International Conference on System Modeling and Advancement in Research Trends, SMART 2019, 2020, pp. 3–6, 9117384

[24] Govindarajan, V., Sivakumar, R., Patil, P.P., Kannan, M., Ramesh, B., Effect of Tungsten Carbide Addition on the Microstructure and Mechanical Behavior of Titanium Matrix Developed by Powder Metallurgy Route, Advances in Materials Science and Engineeringthis link is disabled, 2022, 2022, 2266951

[25] Gupta, A., Singh, R., Nassa, V.K., Sharma, P., Koti, K., Investigating Application and Challenges of Big Data Analytics with Clustering, 2021 International Conference on Advancements in Electrical, Electronics, Communication, Computing and Automation, ICAECA 2021, 2021

[26] H. Kato, M. Billinghurst, R. Blanding, and R. May. Artoolkit, 1999.

[27] J. Irizarry, M. Gheisari, and G. Williams. Infospot: A mobile augmented reality method for accessing building information through a situation awareness approach. J. Autom. Constr., 33:11–23, 2012.

[28] Jain, P., Vijay, S., & Gupta, S. C. (2009). ATM Network QoS Parameters and Performance Analysis using OPNET Simulator. International Journal of Recent Trends in Engineering, 2(4), 228.

[29] Jain, P., Vijay, S., & Gupta, S. C. Performance Analysis & QoS Guarantee in ATM Networks. Global Journal of computer Science and Technology, 131-136.

[30] Jain, S., Choudhury, T., Kumar, V., Kumar, P., Detecting malware and analysing using sandbox evasion, Proceedings of the 2018 International Conference On Communication, Computing and Internet of Things, IC3IoT 2018, 2019, pp. 111–116, 8668149

[31] Kala, S., Nandan, H., Sharma, P., Shadow and weak gravitational lensing of a rotating regular black hole in a non-minimally coupled Einstein-Yang-Mills theory in the presence of plasma, European Physical Journal Plusthis link is disabled, 2022, 137(4), 457

[32] Kandpal, H., Kumar, A., Malik, A., Morphometric analysis and prioritization of sub-watersheds in Bino watershed, Uttarakhand: A remote sensing and GIS perspective, Indian Journal of Ecology, 2018, 45(3), pp. 475–478

[33] Kannan, M., Singh, S., Prasad, R.R., Synthetic Methods to Obtain Calcia-Stabilized Zirconia Powders—A Review, Lecture Notes in Mechanical Engineeringthis link is disabled, 2021, pp. 405–416

[34] Kukreti, A., Painolli, A.K., Rana, N., Where do we stand: factors affecting sustainable development, Procedia Environmental Science, Engineering and Management, Volume 8, Issue 2, 2021, Pages 583-589

[35] Kulkarni, S & Chima, P 2021 'Challenges faced by UK university students due to thcoronavirus crisis in the Higher Education' Preprints.org. https://doi.org/10.20944/preprints202102.0192.v1

[36] Kulkarni, S & Sodre, JR 2017, An Assessment of Lightweight Vehicles to Reduce Greenhouse Gas Emissions with Focus on Magnesium. in Thoughts and Refections on the use of Magnesium. Birmingham City University, pp. 84-91.

[37] Kulkarni, S, Chapman, C & Shah, H 2016 'Computational Fluid Dynamics (CFD) Mesh Independency Study of A Straight Blade Horizontal Axis Tidal Turbine' Preprints.org. https://doi.org/10.20944/preprints201608.0008.v1

[38] Kulkarni, S, Chapman, C, Shah, H & Edwards, D 2018, 'A computational design method for bio-mimicked horizontal axis tidal turbines', International Journal of Building Pathology and Adaptation, vol. 36, no. 2, pp. 188-209. https://doi.org/10.1108/IJBPA-06-2017-0029

[39] Kulkarni, S, Chapman, C, Shah, H, Parn, E & Edwards, DJ 2018, 'Designing an efficient tidal turbine blade through bio-mimicry: A systematic review', Journal of Engineering, Design and Technology, vol. 16, no. 1, pp. 101-124. https://doi.org/10.1108/JEDT-08-2017-0077

[40] Kulkarni, S, Chapman, C, Shah, H, Parn, E & Edwards, DJ 2018, 'Design Study of Horizontal Axis Tidal Turbine Blade', Mindanao Journal of Science and Technology. https://www.open-access.bcu.ac.uk/10089/

[41] Kulkarni, S, Edwards, DJ, Chapman, C, Hosseini, MR & Owusu-Manu, D-G 2019, 'A preliminary mechanical design evaluation of the Wikispeed car: for light-weighting implications', Journal of Engineering, Design and Technology, vol. 17, no. 1, pp. 230-249. https://doi.org/10.1108/JEDT-09-2018-0154

[42] Kulkarni, S, Edwards, DJ, Parn, E, Chapman, C, Aigbavboa, CO & Cornish, R 2018, 'Evaluation Of Vehicle Lightweighting To Reduce Greenhouse Gas Emissions With Focus On Magnesium Substitution', Journal of Engineering, Design and Technology, vol. 16, no. 6, pp. 869-888. https://doi.org/10.1108/JEDT-03-2018-0042

[43] Kulkarni, SS, Wang, L, Golsby, N & Lander, M 2021, 'Fluid-structure interaction based optimisation in tidal turbines: A perspective review', Journal of Ocean Engineering and Science. https://doi.org/10.1016/j.joes.2021.09.017

[44] M. Billinghurst, M., H. Kato, H. and I. Poupyrev, The Magic Book: a transitional AR interface, Computers & Graphics, vol.25:5, pp. 745-753, Elsevier, 2001

[45] M. Tonnis. Data management for augmented reality applications. Master's thesis, Technische Universitat Munchen, 2003.

[46] Malik, A., Kumar, A., Kandpal, H., Morphometric analysis and prioritization of sub-watersheds in a hilly watershed using weighted sum approach, Arabian Journal of Geosciencesthis link is disabled, 2019, 12(4), 118

[47] Metaio. http://www.metaio.com/, 2014.

[48] Mohanakurup, V.K., Baghela, V.S., Kumar, S., Srivastava, P.K., Doohan, N.V., Soni, M., Awa, H., 5G Cognitive Radio Networks Using Reliable Hybrid

Deep Learning Based on Spectrum Sensing, Green Internet of Things in 5G and Beyond, Hindawi, Volume 2022

[49] Mohta, M., Soni, N., Choudhury, T., Kumar, V., Market Analysis on Li-Fi Technology, Proceedings of the 2019 8th International Conference on System Modeling and Advancement in Research Trends, SMART 2019, 2020, pp. 139–143, 9117427

[50] Nagwan, H., Rana, N., Kumar, A., Tewari, P.C., Performance and emission characteristics of diesel engine using Karanja oil methyl ester and Eucalyptus oil fuel blends, Procedia Environmental Science, Engineering and Management, Volume 6, Issue 4, 2019, Pages 553-563

[51] Painoli, A.K., Bansal, R., Singh, R., Kukreti, A., Impact of digital marketing on the buying behavior of youth with special reference to Uttarakhand state, Big Data Analytics for Improved Accuracy, Efficiency, and Decision Making in Digital Marketing, 2021, pp. 162–182

[52] Parag, J. (2011). Hybrid mixed handover for call blocking probability analysis in wireless ATM network. Journal of Engineering and Computer Innovations, 2(4), 68-77.

[53] R. Freitas and P. Campos. Smart: a system of augmented reality for teaching 2nd grade students. In Proceedings of British Computer Society Conference on Human-Computer Interaction, pages 27–30, 2008.

[54] S. J. Henderson and S. K. Feigner. Augmented reality in the psychomotor phase of a procedural task. In Proceedings of ISMAR, pages 191–200, 2011.

[55] S. Julier, Y. Baillot, L. M, and D. Brown. Bars: Battlefield augmented reality system. In NATO Information Systems Technology Panel Symposium on New Information Processing Techniques for Military Systems, 2000.

[56] S. Kosta, A. Aucinas, and P. Hui. Thinkair: Dynamic resource allocation and parallel execution in cloud for mobile code offloading. In Proceedings of INFOCOM, pages 945–953, 2012.

[57] Sharma, C., Singh, M.P., Chaudhary, S., Cost Analysis of Solar-Wind Based Hybrid Renewable Energy System, Lecture Notes in Electrical Engineeringthis link is disabled, 2021, 700, pp. 2733–2746

[58] Sharma, S., Chandra, U., & Jain, P. (2017). A literature survey on automation of test data generation for branch coverage testing using genetic algorithm. International Journal of Computational Intelligence Research, 13(6), 1521-1531.

[59] T. Hollerer, S. Feiner, and D. Hallaway. User interface management techniques for collaborative mobile augmented reality, computers and graphics. Modeling and Simulation Design, 25(5):799–810, 2001.

[60] Varshney, R., Jain, P., & Vijay, S. (2018, September). Massive MIMO Systems In Wireless Communication. In 2018 2nd International Conference on Micro-Electronics and Telecommunication Engineering (ICMETE) (pp. 39-44). IEEE.

[61] W. Holden. http://www.juniperresearch.com/viewpressrelease.php pr=166, 2009.

[62] W. Holden. http://www.juniperresearch.com/viewpressrelease.php pr=232, 2011.

[63] W. Piekarski and B. H. Thomas. Arquake: The outdoor augmented reality gaming system. Communications of the ACM, 45(1):36–38, 2002.

[64] W. Piekarski and H. T. Bruce. An object-oriented software architecture for 3d mixed reality applications. In Proceedings of ISMAR, pages 247–256, 2003.

[65] Z. P. Huang, P. Hui, C. Peylo, and D. Chatzopoulos. Mobile augmented reality survey: A bottom-up approach. Technical report, HKUST Technical Report, 2013 (cite as arXiv:1309.4413).

Chapter - 05

Neural Network (Digital Image Processing) Using Machine Learning

Dr. Sanjay B. Waykar
Mahatma Gandhi Mission's College of Engineering and Technology, India.
Email:sbwaykar@gmail.com

Dr. Anil W. Kale
Mahatma Gandhi Mission's College of Engineering and Technology, India.
Email: anil5474@gmail.com

Dr. Nandkishor P Karlekar
MGMCET Kamothe Navimumbai, India. Email: npkarlekar.mgmcet@gmail.com

Dr. Vaibhav Narawade
Ramrao Adik Institute of Technology, India. Email:vnarawade@gmail.com

INTRODUCTION

Our development "NEURAL NETWORK (Digital Image Processing) "is identified with computerized picture preparing utilizing canny neural organizations framework and furthermore an identifies with picture acknowledgments frameworks for: (a) performing searches of pictures and sharing hunts via web-based media to adapt indexed lists; (b) preparing neural organizations to recognize items; and (c) choosing and buying stock on the web.

Advanced picture preparing regularly includes handling a computerized picture, for instance, from a computerized still picture or computerized video, to determine, recognize, or potentially group specific elements or items in the picture. Example acknowledgment might be applied during the picture preparing to identify a specific article in the picture. Computerized picture preparing with design acknowledgment has been utilized in a wide assortment of utilizations, like facial acknowledgment, identification of land highlights from ethereal photos, vehicle tag assurance, and so on Various sorts of customary AI capacities might be utilized for design acknowledgment, nonetheless, numerous traditional AI capacities are not adjusted or might be hard to adjust for design acknowledgment in advanced picture preparing.

This particular identifies with preparing pictures utilizing profound neural organizations, e.g., convolutional neural organizations. Convolutional neural organizations for the most part incorporate two sorts of neural organization layers, convolutional neural organization layers and completely associated neural organization layers. Convolutional neural organization layers have meager availability, with every hub in a convolutional layer getting input from just a subset of the hubs in the following least neural organization layer. Some convolutional neural organization layers have hubs that offer loads with different hubs in the layer. Hubs in completely associated layers, notwithstanding, get input from every hub in the following most reduced neural organization layer.

In a first angle, the current innovation gives a framework to adapting query items based on particularly produced and saved URLs. In particular, the current framework contains a favored technique for a client to adapt picture looks for an article of product, including:

1. the client transferring a picture to a PC framework in a hunt question;
2. the PC framework utilizing picture acknowledgment programming to discover pictures like the transferred picture in the inquiry;
3. the PC framework showing to the client the pictures that are like the transferred picture, wherein the presentation of pictures is introduced to the client as a website page having a special URL;
4. the client saving the interesting URL;
5. the client sharing the one of a kind URL via web-based media;
6. the client being paid when a subsequent client: (i) sees the novel URL, (ii) prefers the extraordinary URL, (iii) shares the exceptional URL, or (iv) buys the article of product through the remarkable URL. Ideally, the client is paid by the business substance controlling the PC framework, and the sum paid to

the client is determined as a level of the buy made constantly client to a dealer of the article of product.

A benefit of this part of the innovation is that the current methodology makes, saves and offers remarkable URLs for its pursuits. Frameworks presently exist for performing on the web stock looking. Be that as it may, with the current option of novel pursuit URLs added to the ventures, various individuals can perform (and refresh) distinctive indexed lists, with these various clients sharing their own query items with others. Accordingly, different clients of the framework might figure out how to trust or follow the ventures of searchers they are following. This gives a framework wherein clients can best discover the merchandise they are searching for online by trusting the pursuits performed by people having comparable preferences.

In other favored perspectives, the query items depend on inclinations from different clients in a partiality bunch that incorporates the client. Enrollment in the liking gathering can be founded on likenesses in inclinations of buying the article of product. For instance, the inclinations of buying the article of product can remember likenesses for: (I) sum spent to buy the article of product, (ii) the recurrence of buying the article of product, or (iii) the personality of the dealer of the article of product. The benefit of utilizing a fondness bunch is that liking bunches help with advancing query items. In particular, the list items given to one client can be founded on comparative indexed lists given to people who make comparable buys and have comparative preferences.

In favored viewpoints, the picture transferred by the client is a picture from a video, with the client labeling the picture from the video with watchwords. In discretionary parts of the current framework, the query items can be shown as 2D pictures, 3D pictures, or pictures in computer generated experience (for example shown over fanciful or remote foundations) or increased reality (showed over a foundation picture as right now saw by a cell phone camera). In additional discretionary parts of the development, extra query not really settled and showed for the client as the client looks down the page.

In other favored perspectives, the picture search can be iterative with the aftereffects of the inquiry creating results that are taken care of into the following hunt. A particularly iterative pursuit can be performed by: (1) the client seeing the showed pictures, (2) the client choosing one of the showed pictures as a favored picture, (3) the PC framework iteratively refreshing the inquiry question utilizing picture acknowledgment programming to discover pictures like the favored picture, and (4) the PC framework showing to the client the pictures that are like the favored picture. Steps (1) to (4) can be rehashed quite a few times, and the PC framework can show the favored picture along with the pictures that are like the favored picture at every emphasis of the hunt.

Benefits of the iterative quests can incorporate inquiries that are kept up with constantly exceptional (with the latest articles of product being distinguished by one client to help different clients). In a subsequent angle, the current creation gives a framework to choosing 3D articles either for a client to print, or to have others print for the client. In particular, the current framework incorporates a strategy for a client to choose an article of product online for 3D printing, including: (a) the client transferring a picture to a PC framework in a pursuit inquiry; (b) the PC framework utilizing picture acknowledgment programming to discover pictures like the transferred picture in the hunt question; (c) the PC framework showing to the client the pictures that are like the transferred picture; (d) the client choosing one of the showed pictures, in this way choosing an article of product comparing thereto; and (e) the client buying the article of product for 3D printing by: (i) downloading a 3D print model of the article of product and afterward 3D printing the article of product, or (ii) buying the article of product from a seller that 3D prints the article of product. The assurance with regards to whether to buy the 3D article of product from the seller can incorporate choosing the merchant based on: (I) vicinity to the client, or (ii) cost. The PC framework might settle on this choice consequently, or the PC framework may rather show a rundown of merchants, and the client would then be able to choose the seller.

A benefit of this framework is that it utilizes a picture acknowledgment web crawler, instead of just a watchword based web index while choosing the 3D pictures. One more benefit of this strategy is that list items can be immediately refreshed, depending on the situation. In discretionary parts of the innovation, non-3D (i.e.: 2D) pictures are rather looked, ideally to discover articles of product comparing thereto. Also, picture acknowledgment frameworks utilizing neural organizations and AI can be prepared to recognize 3D items dependent on 2D pictures taken at various points or through neural organizations that help with arranging the 3D articles.

Targets

1. The target of the development is to is an advanced picture might be prepared by a gathering of convolutional neural organizations (CNNs) to group objects in the computerized picture and each

CNN, an applicant design and up-and-comer boundaries might be chosen to fabricate a majority of CNNs.

2. The other target of the development is to is the foreordained number of CNNs, each having various qualities for the chose up-and-comer boundaries meet an approval limit a gathering of CNNs might be produced from the foreordained number of CNNs and the expectations from the troupe of CNNs may then be collected to precisely group the items in the advanced picture.
3. The other target of the development is to is Neural Networks is an advanced picture handling normally includes preparing a computerized picture, for instance, from a computerized still picture or advanced video, to learn, distinguish, and characterize specific components or articles in the picture and the example acknowledgment might be applied during the picture preparing to identify a specific item in the picture.
4. The other target of the creation is to is advanced picture handling with design acknowledgment has been utilized in a wide assortment of utilizations, like facial acknowledgment, location of land highlights from aeronautical photos, vehicle tag assurance, and so forth

RUNDOWN

As a general rule, this particular portrays strategies for handling pictures utilizing profound neural organizations. Specific exemplifications of the topic portrayed in this particular can be carried out to acknowledge at least one of the accompanying benefits. By including subnetworks and, specifically, module subnetworks, in a profound neural organization, the profound neural organization can perform better on picture preparing assignments, e.g., object acknowledgment or picture order. Also, profound neural organizations that incorporate module subnetworks can be prepared speedier and more effectively than profound neural organizations that do exclude module subnetworks while keeping up with further developed execution on the picture handling undertakings. The subtleties of at least one encapsulations of the topic of this determination are gone ahead in the going with drawings and the portrayal underneath. Different elements, viewpoints, and benefits of the topic will become evident from the portrayal, the drawings, and the cases.

In certain executions, during preparing, the profound neural organization is expanded with at least one other preparing subnetworks that are taken out after the profound neural organization has been prepared. Each other preparing subnetwork (likewise alluded to as a "side pinnacle") incorporates at least one ordinary neural organization layers, e.g., can incorporate at least one of normal pooling layers, completely associated layers, dropout layers, etc, and a yield layer that is designed to create similar characterizations as the yield layer of the framework. Each other preparing subnetwork is designed to get the yield produced by one of the subnetworks of the profound neural organization, i.e., in corresponding with the subnetwork that as of now gets the subnetwork yield, and interaction the subnetwork yields to create a preparation subnetwork yield for the preparation picture. The preparation subnetwork yield is likewise used to change esteems for the boundaries of the layers in the profound neural organization as a component of the backpropagation preparing strategy. As portrayed above, when the profound neural organization has been prepared, the preparation subnetworks are eliminated.

Epitomes of the topic and the utilitarian tasks portrayed in this determination can be carried out in advanced electronic hardware, in unmistakably epitomized PC programming or firmware, in PC equipment, incorporating the designs unveiled in this detail and their underlying counterparts, or in mixes of at least one of them. Epitomes of the topic portrayed in this determination can be carried out as at least one PC programs, i.e., at least one modules of PC program directions encoded on an unmistakable non passing project transporter for execution by, or to control the activity of, information handling mechanical assembly.

On the other hand, or also, the program guidelines can be encoded on a misleadingly created spread sign, e.g., a machine-produced electrical, optical, or electromagnetic sign, that is produced to encode data for transmission to reasonable recipient mechanical assembly for execution by an information handling contraption. The PC stockpiling medium can be a machine-decipherable capacity gadget, a machine-comprehensible capacity substrate, an irregular or sequential access memory gadget, or a blend of at least one of them.

The expression "information preparing device" incorporates a wide range of mechanical assembly, gadgets, and machines for handling information, including via model a programmable processor, a PC, or different processors or PCs. The device can incorporate unique reason rationale hardware, e.g., a FPGA (field programmable door cluster) or an ASIC (application explicit coordinated circuit). The device can likewise incorporate, notwithstanding equipment, code that establishes an execution climate for the PC program being referred to, e.g., code that comprises processor firmware, a convention stack, an information base administration framework, a working framework, or a blend of at least one of them.

A PC program (which may likewise be alluded to or portrayed as a program, programming, a product application, a module, a product module, a content, or code) can be written in any type of programming language, including aggregated or deciphered dialects, or explanatory or procedural dialects, and it tends to be sent in any structure, including as an independent program or as a module, part, subroutine, or other unit appropriate for use in a figuring climate.

A PC program may, however need not, relate to a record in a document framework. A program can be put away in a piece of a record that holds different projects or information, e.g., at least one contents put away in a markup language archive, in a solitary document devoted to the program being referred to, or in various composed records, e.g., records that store at least one modules, sub projects, or divides of code. A PC program can be conveyed to be executed on one PC or on different PCs that are situated at one site or appropriated across various destinations and interconnected by a correspondence organization.

The cycles and rationale streams depicted in this particular can be performed by at least one programmable PCs executing at least one PC projects to perform capacities by working on input information and producing yield. The cycles and rationale streams can likewise be performed by, and device can likewise be executed as, unique reason rationale hardware, e.g., a FPGA (field programmable door exhibit) or an ASIC (application explicit incorporated circuit).

PCs reasonable for the execution of a PC program incorporate, via model, can be founded on broad or particular reason microchips or both, or some other sort of focal preparing unit. For the most part, a focal preparing unit will get guidelines and information from a read just memory or an arbitrary access memory or both. The fundamental components of a PC are a focal handling unit for performing or executing guidelines and at least one memory gadgets for putting away directions and information. By and large, a PC will likewise incorporate, or be operatively coupled to get information from or move information to, or both, at least one mass stockpiling gadgets for putting away information, e.g., attractive, magneto optical circles, or optical plates. Nonetheless, a PC need not have such gadgets.

Also, a PC can be implanted in another gadget, e.g., a cell phone, an individual computerized partner (PDA), a versatile sound or video player, a game control center, a Global Positioning System (GPS) recipient, or a convenient stockpiling gadget, e.g., an all-inclusive sequential transport (USB) streak drive, to give some examples. PC comprehensible media reasonable for putting away PC program directions and information incorporate all types of nonvolatile memory, media and memory gadgets, including via model semiconductor memory gadgets, e.g., EPROM, EEPROM, and blaze memory gadgets; attractive plates, e.g., inside hard circles or removable circles; magneto optical circles; and CD ROM and DVD-ROM plates. The processor and the memory can be enhanced by, or consolidated in, unique reason rationale hardware.

To furnish for collaboration with a client, encapsulations of the topic depicted in this detail can be executed on a PC having a presentation gadget, e.g., a CRT (cathode beam cylinder) or LCD (fluid gem show) screen, for showing data to the client and a console and a pointing gadget, e.g., a mouse or a trackball, by which the client can give contribution to the PC. Different sorts of gadgets can be utilized to furnish for association with a client also; for instance, criticism gave to the client can be any type of tangible input, e.g., visual input, hearable input, or material input; and contribution from the client can be gotten in any structure, including acoustic, discourse, or material information. What's more, a PC can cooperate with a client by sending archives to and getting reports from a gadget that is utilized by the client; for instance, by sending website pages to an internet browser on a client's customer gadget in light of solicitations got from the internet browser.

Encapsulations of the topic depicted in this detail can be executed in a registering framework that incorporates a back end part, e.g., as an information worker, or that incorporates a middleware part, e.g., an application worker, or that incorporates a front end part, e.g., a customer PC having a graphical UI or a Web program through which a client can collaborate with an execution of the topic portrayed in this particular, or any mix of at least one such back end, middleware, or front end parts. The parts of the framework can be interconnected by any structure or vehicle of computerized information correspondence, e.g., a correspondence organization. Instances of correspondence networks incorporate a neighborhood ("LAN") and a wide region network ("WAN"), e.g., the Internet. The processing framework can incorporate customers and workers. A customer and worker are by and large remote from one another and commonly collaborate through a correspondence organization. The relationship of customer and worker emerges by prudence of PC programs running on the particular PCs and having a customer worker relationship to one another.

While this determination contains numerous particular execution subtleties, these ought not be interpreted as restrictions on the extent of any creation or of what might be asserted, yet rather as portrayals of elements that might be explicit to specific exemplifications of specific developments. Certain components that are depicted in this determination with regards to isolate exemplifications can likewise be executed in blend in a solitary encapsulation.

Then again, different elements that are portrayed with regards to a solitary encapsulation can likewise be carried out in numerous epitomes independently or in any reasonable subcombinations. Additionally, in spite of the fact that provisions might be depicted above as acting in specific blends and surprisingly at first asserted thusly, at least one elements from a guaranteed mix can at times be extracted from the mix, and the guaranteed mix might be coordinated to a subcombinations or variety of a subcombinations.

Also, while activities are portrayed in the drawings in a specific request, this ought not be perceived as necessitating that such tasks be acted in the specific request shown or in consecutive request, or that all delineated tasks be performed, to accomplish beneficial outcomes. In specific conditions, performing multiple tasks and equal handling might be favorable. In addition, the division of different framework modules and parts in the encapsulations depicted above ought not be perceived as requiring such partition in all exemplifications, and it ought to be perceived that the portrayed program parts and frameworks can by and large be incorporated together in a solitary programming item or bundled into numerous product items.

METHOD

Our Invention" NEURAL NETWORK (Digital Image Processing)" is a computerized picture might be prepared by an outfit of convolutional neural organizations (CNNs) to group objects in the advanced picture and each CNN, a competitor engineering and up-and-comer boundaries might be chosen to fabricate a majority of CNNs. The created innovation likewise incorporates the foreordained number of CNNs, each having various qualities for the chose up-and-comer boundaries meet an approval limit an outfit of CNNs might be produced from the foreordained number of CNNs and the forecasts from the group of CNNs may then be collected to precisely arrange the articles in the advanced picture. The DIP-Intelligent Neural Networks is an advanced picture preparing regularly includes handling a computerized picture, for instance, from a computerized still picture or computerized video, to find out, recognize, and characterize specific elements or items in the picture and the example acknowledgment might be applied during the picture handling to distinguish a specific article in the picture. The created innovation additionally an advanced picture preparing with design acknowledgment has been utilized in a wide assortment of utilizations, like facial acknowledgment, location of land highlights from ethereal photos, vehicle tag assurance.

DIAGRAM

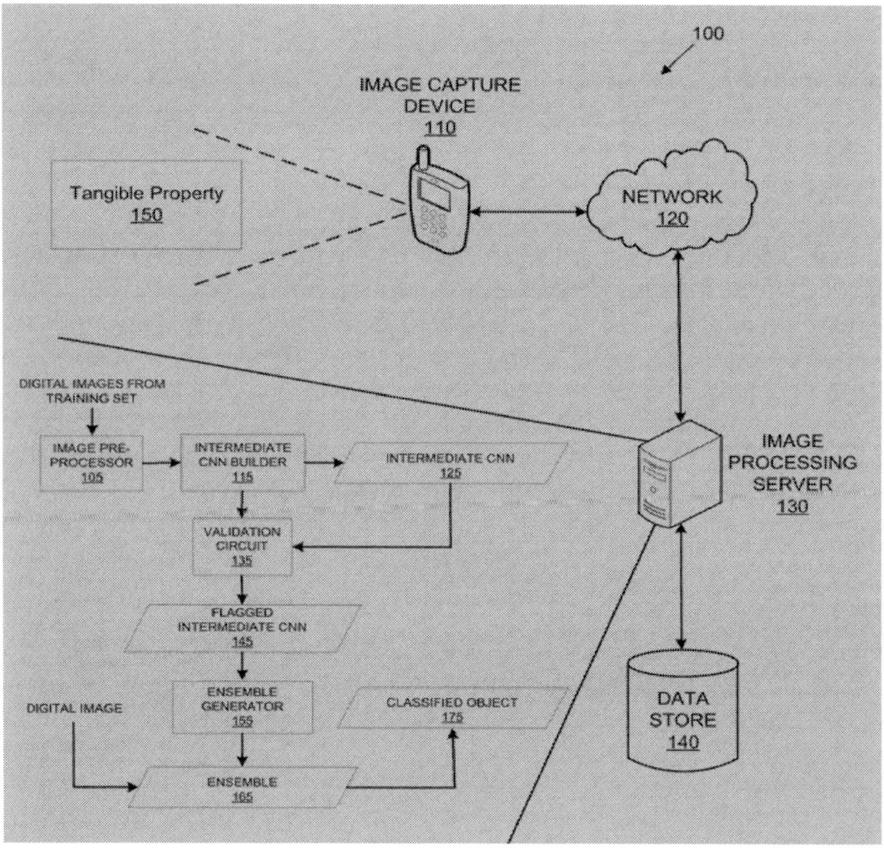

Fig. 1: Shows a system diagram of an image processing system, according to an example of the present disclosure;

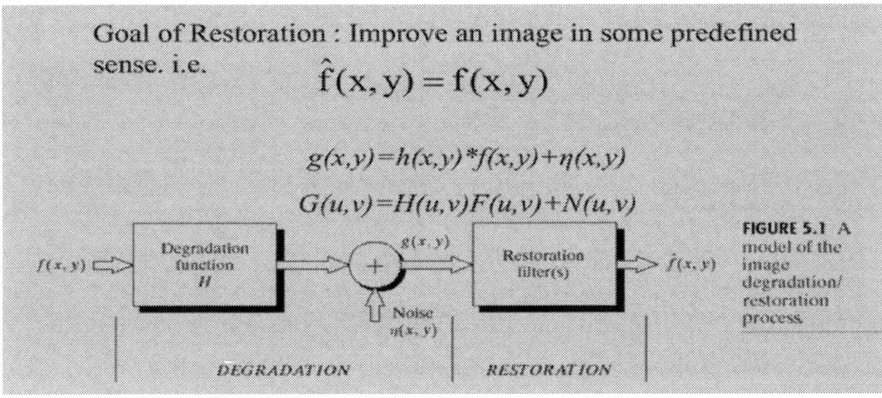

Fig. 2: Shows classification categories that indicate the extent of damage to property, according to an example of the present disclosure;

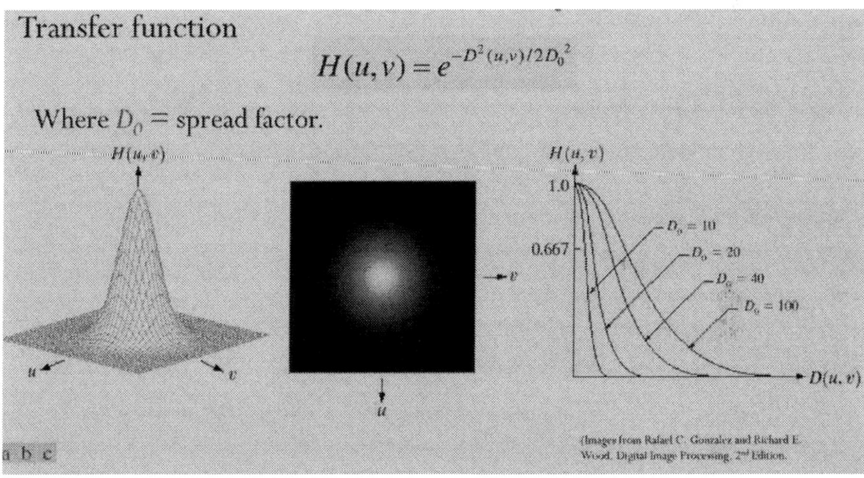

Fig. 3: Shows a data store of an image processing server, according to an example of the present disclosure.

DESCRIPTION

For straightforwardness and illustrative purposes, the current revelation is depicted by alluding chiefly to a model thereof. In the accompanying portrayal, various explicit subtleties are gone ahead to give an intensive comprehension of the current revelation. It will be promptly evident notwithstanding, that the current exposure might be drilled without restriction to these particular subtleties.

In different occasions, a few techniques and designs have not been portrayed exhaustively so as not to pointlessly darken the current divulgence. As utilized in this, the expressions "a" and "an" are planned to signify no less than one of a specific component, the expression "incorporates" signifies incorporates yet not restricted to, the expression "counting" signifies including yet not restricted to, and the expression "in light of" signifies put together basically partially with respect to.

A picture handling framework, as per a model, forms and prepares a gathering of profound learning models, for example, convolutional neural organizations (CNNs), to precisely and naturally perform picture handling to distinguish specific credits of items in a computerized picture, and to arrange the articles as per the recognized characteristics. CNNs, notwithstanding, incorporate numerous practical parts, which make it hard to decide the fundamental organization design that performs precisely to identify and order specific elements of pictures applicable for an issue close by.

Moreover, every part of the CNN ordinarily has a large number of boundaries related with it. The particular upsides of those boundaries fundamental for an effective and exact picture arrangement are not known deduced with no utilization of a powerful picture handling framework. The picture preparing framework, consequently, gives a technique to building and tweaking CNNs demonstrated to yield an exact grouping of a picture.

Through an iterative cycle, an up-and-comer design and up-and-comer boundaries for CNN might be chosen to fabricate, train, and advance a CNN. For instance, the iterative cycle might incorporate choosing the up-and-comer design from a majority of up-and-comer models and approving a bunch of applicant boundaries for the chose up-and-comer engineering. The applicant design might incorporate various convolution layers and subsampling layers and a classifier type. The competitor boundaries might incorporate a learning rate, a bunch size, a most extreme number of preparing ages, an info picture size, various component maps at each layer of the CNN, a convolutional channel size, a sub-examining pool size, various secret layers, various units in each secret layer, a chose classifier calculation, and various yield.

The iterative process may include building an intermediate CNN using the training set and evaluating the performance of the intermediate CNN on a validation set. The evaluation, for instance, determines whether the intermediate CNN meets a validation threshold such as less than a 20% error rate. The iterative process is repeated until a predetermined number of intermediate CNNs (e.g., 25) meet the validation threshold. According to an example, each intermediate CNN has different values for the selected candidate parameters.

An ensemble of the most accurate intermediate CNNs is then generated from the predetermined number of intermediate CNNs. The ensemble for example may be the top 5 most accurate intermediate CNNs. The next step may include selecting an ensemble algorithm to aggregate and/or combine the predictions of each intermediate CNN in the ensemble to form an ensemble prediction. The prediction of each intermediate CNN in the ensemble may then be used to classify an image or an object in the image.

The technical benefits and advantages of the disclosed examples include providing an advanced deep learning architecture that exhibits superior classification accuracy to assess property damage and an iterative image processing system that determines that advanced deep learning architecture. A CNN generated by the image processing system through an iterative process is easier to train than other regular, feed-forward neural networks and has fewer parameters to estimate, making it a more efficient architecture to use to assess property damage.

According to an example, a CNN generated by the image processing system may be used for classifying the extent of damage to a property that is captured in a digital image. Damage may refer to any kind of injury or harm that impairs the appearance of the property. An image or digital image may include both a still image and a moving image (e.g., video). The property may be any tangible object including, but not limited to, a house, furniture, clothing, vehicle equipment, land, computing device, toy, etc. In an example where an insured customer has accidental damage to tangible property, the insured customer may document the damage to the damaged property by taking digital photographs with a smart phone and/or camera.

The digital images of the damaged property may then be fed into the image processing system. The image processing system may automatically classify the damaged property based on amount of damage determined from the image processing of the received digital images. In this example, the image processing system provides a machine vision method and apparatus to automatically detect the extent of damage to the property as captured in digital images.

According to an example, the image processing system generates an ensemble model (e.g., including multiple optimized CNNs) to classify an image or an object in the image with improved accuracy. In an example, the image processing system which used the ensemble model yielded an accuracy of nearly 90% on the images in the validation set. As discussed above, according to an example, the image processing system may be used for classifying the extent of damage to property captured in an image. However, the image processing system may be used for substantially any application to classify features in a digital image into predefined categories.

With reference to FIG. 1, there is shown a system diagram of an image processing system 100, according to an example of the present disclosure. It should be understood that the system 100 may include additional components and that one or more of the components described herein may be removed and/or modified without departing from a scope of the system 100. The system 100 may include at least one image capture device 110, a communications network 120, an image processing server 130, and a data store 140.

The image capture device 110 may communicate with the image processing server 130 via the communications network 120. The image capture device 110 may be any computing device that includes a camera such as, but not limited to, a smartphone, a computing tablet, a laptop computer, a desktop computer, or any wearable computing device. According to an example, the image capture device 110 may capture an image of a tangible property 150 and send the image of the tangible property 150 to the image processing server 130 to automatically classify the extent of damage to the tangible property 150. classes. Likewise, a pre-preparing convention may likewise be chosen to improve specific substance in the pictures for the chose up-and-comer design and chose competitor boundaries.

The interchanges network 120 might incorporate neighborhood (LANs) and wide region organizations (WANs), like the Internet. The correspondences network 120 might incorporate sign bearing mediums that might be constrained by programming, applications and additionally rationale. The interchanges network 120 might incorporate a mix of organization components to help information correspondence administrations. For instance, the interchanges network 120 might interface the picture catch gadget 110 to the picture preparing worker 130 using an actual association, for example, copper link, coaxial link, and fiber link, or through remote innovation like radio, microwave, or satellite.

The picture handling worker 130, for instance, may get advanced pictures from a preparation set at a picture pre-processor 105. The picture pre-processor might trim and improve specific substance in the pictures from the preparation set to enter into the moderate CNN developer 115. The middle of the road CNN developer 115 may choose different structures and boundaries to prepare a moderate CNN 125. The moderate CNN 125 might be then being assessed on an approval set that is created by the approval circuit 135. The approval circuit 135 may decide if to hail the halfway CNN 125 as meeting an assigned approval limit. In the event that the middle CNN 125 doesn't meet the approval edge, the transitional CNN isn't hailed and keeps on being prepared on the computerized pictures from the preparation set by the halfway CNN developer 115

Nonetheless, if the middle CNN 125 meets the approval limit, the transitional CNN 125 is presently a hailed halfway CNN 145. Thus, the hailed middle CNN 145 is qualified to be chosen as a feature of an outfit of streamlined CNNs that is created by the group generator 155. The group generator 155, for instance, may make a gathering 165 of advanced CNNs. The expectations totaled from the group 165 might be utilized to precisely arrange objects 175 from an inputted advanced picture. The preparing elements of the picture handling worker 130 are additionally itemized beneath .As per a model, the picture preparing worker 130 may get a picture of the unmistakable property 150 and consequently group a degree of harm to the substantial property 150 utilizing CNNs to perceive and arrange the harm in the picture of the substantial property 150. As per a model, the picture preparing worker 130 may group the degree of harm to the substantial property 150 into different foreordained arrangement classes 200, for example, yet not restricted to, flawless, harmed, and seriously harmed or added up to as outlined in FIG. 2.

OUTCOME

Our Invention "NEURAL NETWORK (Digital Image Processing)" is a computerized picture might be prepared by a gathering of convolutional neural organizations (CNNs) to order protests in the advanced picture and each CNN, an up-and-comer engineering and up-and-comer boundaries might be chosen to assemble a majority of CNNs. The imagined innovation likewise incorporates the foreordained number of CNNs, each having various qualities for the chose competitor boundaries meet an approval edge a group of CNNs might be produced from the foreordained number of CNNs and the expectations from the gathering of CNNs may then be accumulated to precisely characterize the articles in the computerized picture. The DIP-Intelligent Neural Networks is an advanced picture handling normally includes preparing a computerized picture, for instance, from a computerized still picture or advanced video, to learn, recognize, and characterize specific provisions or articles in the picture and the example acknowledgment might be applied during the picture handling to identify a specific item in the picture. The created innovation additionally an advanced picture handling with design acknowledgment has been utilized in a wide assortment of uses, like facial acknowledgment, discovery of land highlights from elevated photos, vehicle tag assurance, and so on

1. According to claim1# the creation is to is a computerized picture might be handled by a gathering of convolutional neural organizations (CNNs) to arrange objects in the advanced picture and each CNN, a competitor engineering and up-and-comer boundaries might be chosen to fabricate a majority of CNNs.
2. According to claim1,2# the creation is to is the foreordained number of CNNs, each having various qualities for the chose competitor boundaries meet an approval limit a gathering of CNNs might be produced from the foreordained number of CNNs and the expectations from the troupe of CNNs may then be amassed to precisely characterize the articles in the advanced picture.
3. According to claim1,2,3# the creation is to is a Neural Networks is a computerized picture handling normally includes preparing an advanced picture, for instance, from an advanced still picture or computerized video, to learn, identify, and group specific elements or articles in the picture and the example acknowledgment might be applied during the picture preparing to distinguish a specific item in the picture.
4. According to claim1,2,4# the development is to is an advanced picture preparing with design acknowledgment has been utilized in a wide assortment of uses, like facial acknowledgment, identification of land highlights from airborne photos, vehicle tag assurance.

REFERENCES

[1] Abhigna, B.S., Chakraborty, S., Kumar, V., Medha: A collaborative E-concept framework, Proceedings of the 2018 International Conference On Communication, Computing and Internet of Things, IC3IoT 2018, 2019, pp. 117–122, 8668127

[2] Abidin, S., Vadi, V.K., Rana, A., On Confidentiality, Integrity, Authenticity, and Freshness (CIAF) in WSN, Advances in Computer, Communication and Computational Sciences, 2020, pp 87-97

[3] ahay, A., Choudhury, T., Kumar, V., Hierarchical Evolution of Digital Games, Proceedings of 2019 International Conference on Computational Intelligence and Knowledge Economy, ICCIKE 2019, 2019, pp. 381–386, 9004378

[4] Area: Communication, Frequency modulation, Networks, wireless communication

[5] Area: Communication, Frequency modulation, Networks, wireless communication

[6] Area: Communication, Frequency modulation, Networks, wireless communication

[7] Area: Communication, Frequency modulation, Networks, wireless communication

[8] Arora, M., Kumar, V., Dave, M., Hybrid cloud infrastructure for load balancing and scaling with vm path optimization using ant colony optimization, Journal of Advanced Research in Dynamical and Control Systems, 2020, 12(5 Special Issue), pp. 660–674

[9] Arora, M., Kumar, V., Dave, M., Task scheduling in cloud infrastructure using optimization technique genetic algorithm, Proceedings of the World Conference on Smart Trends in Systems, Security and Sustainability, WS4 2020, 2020, pp. 788–793, 9210303

[10] Bansal, R., Gupta, A., Singh, R., Nassa, V.K., Role and impact of digital technologies in E-learning amidst COVID-19 pandemic, Proceedings - 2021 4th International Conference on Computational Intelligence and Communication Technologies, CCICT 2021, 2021, pp. 194–202

[11] Bansal, R., Obaid, A.J., Gupta, A., Singh, R., Pramanik, S., Impact of Big Data on Digital Transformation in 5G Era, Journal of Physics: Conference Seriesthis link is disabled, 2021, 1963(1), 012170

[12] Chauhan, S.S., Khare, A.K., Kinematic analysis of the ABB IRB 1520 industrial robot using roboanalyzer software, Evergreenthis link is disabled, 2020, 7(4), pp. 510–518

[13] Chauhan, S.S., Khare, A.K., Method of Obtaining Planar State of Stress Using Mohr's Circle—Some Typical Cases, Lecture Notes in Mechanical Engineering, 2022, pp. 33–38

[14] Goel, S., Tripathi, A., Choudhury, T., Kumar, V. Parkinson's Disease Detection using Soft Computing Technique, Proceedings of the 2019 8th International Conference on System Modeling and Advancement in Research Trends, SMART 2019, 2020, pp. 3–6, 9117384

[15] Govindarajan, V., Sivakumar, R., Patil, P.P., Kannan, M., Ramesh, B., Effect of Tungsten Carbide Addition on the Microstructure and Mechanical Behavior of Titanium Matrix Developed by Powder Metallurgy Route, Advances in Materials Science and Engineeringthis link is disabled, 2022, 2022, 2266951

[16] Gupta, A., Singh, R., Nassa, V.K., Sharma, P., Koti, K., Investigating Application and Challenges of Big Data Analytics with Clustering, 2021 International Conference on Advancements in Electrical, Electronics, Communication, Computing and Automation, ICAECA 2021, 2021

[17] https://nanonets.com/blog/machine-learning-image-processing/

[18] https://patents.google.com/patent/CA2921672A1

[19] https://towardsdatascience.com/convolution-neural-network-for-image-processing-using-keras-dc3429056306

[20] https://www.isprs.org/proceedings/XXXVII/congress/3b_pdf/109.pdf

[21] https://www.researchgate.net/publication/311668047_Image_Processing_Using_Artificial_Neural_Networks

[22] https://www.vision-systems.com/boards-software/article/16736492/image-processing-and-neural-networks-classify-complex-defects

[23] Jain, P., Vijay, S., & Gupta, S. C. (2009). ATM Network QoS Parameters and Performance Analysis using OPNET Simulator. International Journal of Recent Trends in Engineering, 2(4), 228.

[24] Jain, P., Vijay, S., & Gupta, S. C. Performance Analysis & QoS Guarantee in ATM Networks. Global Journal of computer Science and Technology, 131-136.

[25] Jain, S., Choudhury, T., Kumar, V., Kumar, P., Detecting malware and analysing using sandbox evasion, Proceedings of the 2018 International Conference On Communication, Computing and Internet of Things, IC3IoT 2018, 2019, pp. 111–116, 8668149

[26] Kala, S., Nandan, H., Sharma, P., Shadow and weak gravitational lensing of a rotating regular black hole in a non-minimally coupled Einstein-Yang-Mills theory in the presence of plasma, European Physical Journal Plusthis link is disabled, 2022, 137(4), 457

[27] Kandpal, H., Kumar, A., Malik, A., Morphometric analysis and prioritization of sub-watersheds in Bino watershed, Uttarakhand: A remote sensing and GIS perspective, Indian Journal of Ecology, 2018, 45(3), pp. 475–478

[28] Kannan, M., Singh, S., Prasad, R.R., Synthetic Methods to Obtain Calcia-Stabilized Zirconia Powders—A Review, Lecture Notes in Mechanical Engineeringthis link is disabled, 2021, pp. 405–416

[29] Kukreti, A., Painolli, A.K., Rana, N., Where do we stand: factors affecting sustainable development, Procedia Environmental Science, Engineering and Management, Volume 8, Issue 2, 2021, Pages 583-589

[30] Kulkarni, S & Chima, P 2021 'Challenges faced by UK university students due to thcoronavirus crisis in the Higher Education' Preprints.org. https://doi.org/10.20944/preprints202102.0192.v1

[31] Kulkarni, S & Sodre, JR 2017, An Assessment of Lightweight Vehicles to Reduce Greenhouse Gas Emissions with Focus on Magnesium. in Thoughts and Refections on the use of Magnesium. Birmingham City University, pp. 84-91.

[32] Kulkarni, S, Chapman, C & Shah, H 2016 'Computational Fluid Dynamics (CFD) Mesh Independency Study of A Straight Blade Horizontal Axis Tidal Turbine' Preprints.org. https://doi.org/10.20944/preprints201608.0008.v1

[33] Kulkarni, S, Chapman, C, Shah, H & Edwards, D 2018, 'A computational design method for bio-mimicked horizontal axis tidal turbines', International Journal of Building Pathology and Adaptation, vol. 36, no. 2, pp. 188-209. https://doi.org/10.1108/IJBPA-06-2017-0029

[34] Kulkarni, S, Chapman, C, Shah, H, Parn, E & Edwards, DJ 2018, 'Designing an efficient tidal turbine blade through bio-mimicry: A systematic review', Journal of Engineering, Design and Technology, vol. 16, no. 1, pp. 101-124. https://doi.org/10.1108/JEDT-08-2017-0077

[35] Kulkarni, S, Chapman, C, Shah, H, Parn, E & Edwards, DJ 2018, 'Design Study of Horizontal Axis Tidal Turbine Blade', Mindanao Journal of Science and Technology. https://www.open-access.bcu.ac.uk/10089/

[36] Kulkarni, S, Edwards, DJ, Chapman, C, Hosseini, MR & Owusu-Manu, D-G 2019, 'A preliminary mechanical design evaluation of the Wikispeed car: for light-weighting implications', Journal of Engineering, Design and Technology, vol. 17, no. 1, pp. 230-249. https://doi.org/10.1108/JEDT-09-2018-0154

[37] Kulkarni, S, Edwards, DJ, Parn, E, Chapman, C, Aigbavboa, CO & Cornish, R 2018, 'Evaluation Of Vehicle Lightweighting To Reduce Greenhouse Gas Emissions With Focus On Magnesium Substitution', Journal of Engineering, Design and Technology, vol. 16, no. 6, pp. 869-888. https://doi.org/10.1108/JEDT-03-2018-0042

[38] Kulkarni, SS, Wang, L, Golsby, N & Lander, M 2021, 'Fluid-structure interaction based optimisation in tidal turbines: A perspective review', Journal of Ocean Engineering and Science. https://doi.org/10.1016/j.joes.2021.09.017

[39] Malik, A., Kumar, A., Kandpal, H., Morphometric analysis and prioritization of sub-watersheds in a hilly watershed using weighted sum approach, Arabian Journal of Geosciencesthis link is disabled, 2019, 12(4), 118

[40] Mohanakurup, V.K., Baghela, V.S., Kumar, S., Srivastava, P.K., Doohan, N.V., Soni, M., Awa, H., 5G Cognitive Radio Networks Using Reliable Hybrid Deep Learning Based on Spectrum Sensing, Green Internet of Things in 5G and Beyond, Hindawi, Volume 2022

[41] Mohta, M., Soni, N., Choudhury, T., Kumar, V., Market Analysis on Li-Fi Technology, Proceedings of the 2019 8th International Conference on System Modeling and Advancement in Research Trends, SMART 2019, 2020, pp. 139–143, 9117427

[42] Nagwan, H., Rana, N., Kumar, A., Tewari, P.C., Performance and emission characteristics of diesel engine using Karanja oil methyl ester and Eucalyptus oil fuel blends, Procedia Environmental Science, Engineering and Management, Volume 6, Issue 4, 2019, Pages 553-563

[43] Painoli, A.K., Bansal, R., Singh, R., Kukreti, A., Impact of digital marketing on the buying behavior of youth with special reference to Uttarakhand state, Big Data Analytics for Improved Accuracy, Efficiency, and Decision Making in Digital Marketing, 2021, pp. 162–182.

[44] Parag, J. (2011). Hybrid mixed handover for call blocking probability analysis in wireless ATM network. Journal of Engineering and Computer Innovations, 2(4), 68-77.

[45] Sharma, C., Singh, M.P., Chaudhary, S., Cost Analysis of Solar-Wind Based Hybrid Renewable Energy System, Lecture Notes in Electrical Engineeringthis link is disabled, 2021, 700, pp. 2733–2746.

[46] Sharma, S., Chandra, U., & Jain, P. (2017). A literature survey on automation of test data generation for branch coverage testing using genetic algorithm. International Journal of Computational Intelligence Research, 13(6), 1521-1531.

[47] Varshney, R., Jain, P., & Vijay, S. (2018, September). Massive MIMO Systems In Wireless Communication. In 2018 2nd International Conference on Micro-Electronics and Telecommunication Engineering (ICMETE) (pp. 39-44). IEEE.

CHAPTER - 06

Big Data and Cloud Computing Using AI

Dr. Anil W. Kale

Mahatma Gandhi Mission's College of Engineering and Technology, India.
Email: anil5474@gmail.com

Dr. Nandkishor P Karlekar

MGMCET Kamothe Navimumbai, India. Email: npkarlekar.mgmcet@gmail.com

Dr. Vaibhav Narawadc

Ramrao Adik Institute of Technology, India. Email:vnarawade@gmail.com

Dr. Sanjay B. Waykar

Mahatma Gandhi Mission's College of Engineering and Technology, India.
Email:sbwaykar@gmail.com

INTRODUCTION

The framework, techniques, and PC program items are introduced for performing cloud information execution and investigation utilizing AI. The imagined innovation alos unaided and regulated learning module is arranged to gather an unstructured informational collection into numerous renditions of a coordinated informational collection.

The IML-cloud information execution is a regulated learning module is designed to produce more the one AI troupe's dependent on every form of different adaptations of a coordinated informational index and to figure out which AI outfit displays a most elevated prescient exhibition.

The IML-cloud information execution is a technique for performing cloud information execution and investigation utilizing AI and furthermore a strategy, incorporates removing information from at least one cloud information sources. The created innovation likewise incorporates stacking cloud information into an unstructured informational collection having an unstructured configuration and a strategy in certain incorporates gathering an unstructured informational collection into a coordinated informational index having an organized arrangement and furthermore a technique incorporates producing at least one learned capacities dependent on a coordinated informational collection.

Huge Data is an information investigation approach empowered by another age of innovations and design which support high-speed information catch, stockpiling, and examination. Information sources reach out past the conventional corporate data set to incorporate email, cell phone yield, sensor-created information, and web-based media yield. Information are at this point not limited to organized data set records yet incorporate unstructured information – information having no standard designing.

Large Data requires colossal measures of extra room. While the cost of capacity kept on declining, the assets expected to use large information can in any case present monetary troubles for little to medium estimated organizations. A normal enormous information stockpiling and investigation foundation will be founded on bunched network-appended capacity (NAS). Bunched NAS framework requires setup of a few NAS "units" with every NAS "case" included a few stockpiling gadgets associated with a NAS gadget. The series of NAS gadgets are then interconnected to permit enormous sharing and looking of information.

Distributed computing is an incredibly effective worldview of administration arranged registering, and has reformed the manner in which figuring foundation is disconnected and utilized. Three most well-known cloud ideal models include: Infrastructure as a Service (IaaS), Platform as a Service (PaaS), and Software as a Service (SaaS). The idea anyway can likewise be stretched out to Database as a Service or Storage as a Service. Flexibility, pay-scrutinize, low forthright speculation, low an ideal opportunity to market, and move of dangers are a portion of the major empowering highlights that make distributed computing a general worldview for sending novel applications which were not financially plausible in a customary endeavor foundation setting.

Versatile information base administration frameworks, both for update serious application jobs, just as choice emotionally supportive networks—are subsequently a basic piece of the cloud foundation. Versatile and appropriated information the executives has been the vision of the data set exploration local area for over thirty years. Much examination has zeroed in on planning versatile frameworks for both update serious jobs just as impromptu investigation responsibilities. Introductory plans incorporate appropriated data sets for update concentrated responsibilities and equal information base frameworks for logical jobs.

Equal information bases developed past model frameworks to enormous business frameworks, yet appropriated data set frameworks were not extremely effective and were rarely popularized – rather different specially appointed ways to deal with scaling were utilized. Changes in the information access examples of utilizations and the need to scale out to a large number of ware machines prompted the introduction of another class of frameworks alluded to as Key-Value stores which are presently being broadly embraced by different ventures. In the space of information investigation, the MapReduce worldview and its open-source execution Hadoop has additionally seen far and wide reception in industry and the scholarly community the same. Arrangements have additionally been proposed to further develop Hadoop based frameworks as far as convenience and execution.

Business knowledge (BI) may incorporate preparing and examination of information for business purposes. Organizations regularly amass a lot of information, with various information made for various purposes and by various sources.

Since conceivably related information across a business element might have distinctive designing and as a rule isn't recognized or filed as being connected, business openings might be missed. Further, manual area and association of related information can be tedious and wrong. Regardless of whether parts of information area and additionally association might be computerized, a human commonly surveys the information, making loose manual approximations and presumptions.

OBJECTIVES

1. The goal of the creation is to a framework technique, and PC program items are introduced for performing cloud information execution and investigation utilizing AI.

2. The other goal of the creation is to an unaided and regulated learning module is arranged to collect an unstructured informational index into different forms of a coordinated informational collection.
3. The other goal of the creation is to a cloud information execution is a directed learning module is designed to produce more the one AI gatherings dependent on every rendition of different forms of a coordinated informational collection and to figure out which AI outfit shows a most elevated prescient exhibition.
4. The other goal of the creation is to a will be a technique for performing cloud information execution and investigation utilizing AI and furthermore a strategy, incorporates extricating information from at least one cloud information sources?
5. The other target of the development is to a stacking cloud information into an unstructured informational collection having an unstructured configuration and a strategy in certain incorporates gathering an unstructured informational collection into a coordinated informational collection having an organized organization and furthermore a technique incorporates creating at least one learned capacities dependent on a coordinated informational collection.

SUMMARY

A contraption is introduced for performi The new huge information worldview happens when the size of the information very still or moving powers the administration of the information to be a huge driver in the plan of the framework engineering. On a very basic level the large information worldview addresses a change in information framework models from solid frameworks with vertical scaling (quicker processors or circles) into an evenly scaled framework that coordinates an approximately coupled arrangement of assets.

This shift happened 20-a few years' prior in the reproduction local area when the logical recreations started utilizing hugely equal preparing (MPP) frameworks. In various blends of parting the code and information across autonomous processors, computational researchers had the option to enormously broaden their reenactment abilities. This obviously presented various entanglements in such regions as message passing, information development, and idleness in the consistency across assets, load adjusting, and framework failures while looking out for different assets to finish their jobs.

Similarly, the large information worldview addresses this equivalent shift, again utilizing various components to circulate code and information across inexactly coupled assets to give the scaling in information dealing with that is expected to coordinate with the scaling in the information. The motivation behind putting away and recovering a lot of information is to perform investigation that delivers extra information about the information. Previously, the investigation was by and large cultivated on an arbitrary example of the information. ng information investigation utilizing AI. In one epitome, a concentrate module is arranged to separate information from at least one organized information sources.

A heap module, in a further exemplification, is arranged to stack information into an unstructured informational index. An unaided learning module, in specific encapsulations, is arranged to collect an unstructured informational index into a coordinated informational collection utilizing a majority of solo learning methods.

One more mechanical assembly for performing information investigation utilizing AI is introduced. In one exemplification, a solo learning module is arranged to gather an unstructured informational collection into numerous renditions of a coordinated informational collection. A regulated learning module, in specific encapsulations, is arranged to create at least one AI outfits dependent on every rendition of numerous variants of a coordinated informational index and to figure out which AI group shows a most noteworthy prescient presentation.

A strategy is introduced for performing information examination utilizing AI. A strategy, in one encapsulation, incorporates removing information from at least one information sources. In a further epitome, a technique incorporates stacking information into an unstructured informational collection having an unstructured arrangement. A technique, in specific exemplifications, incorporates collecting an unstructured informational index into a coordinated informational index having an organized configuration. In another encapsulation, a technique incorporates producing at least one learned capacities dependent on a coordinated informational index.

DIAGRAM

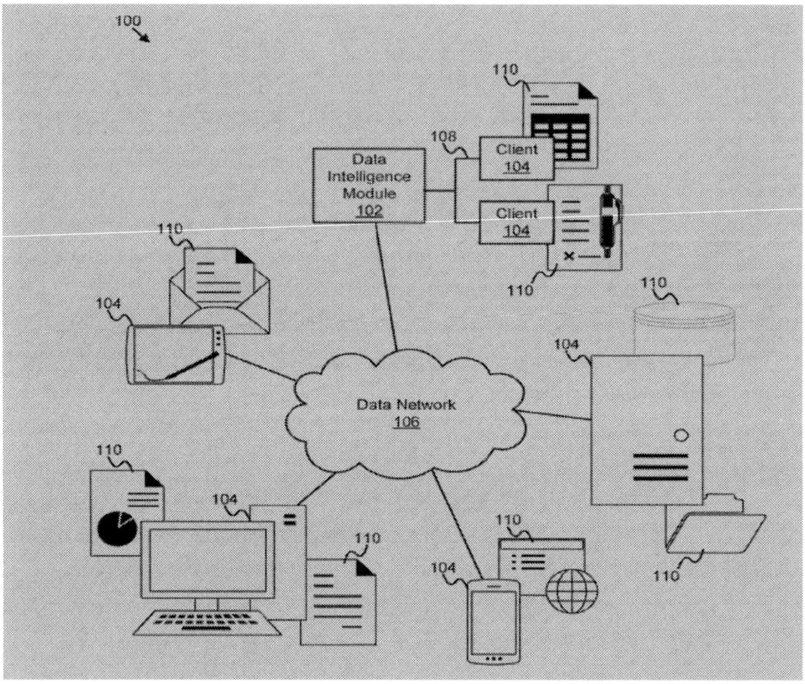

Fig. 1: is a schematic block diagram illustrating one embodiment of a system for data intelligence;

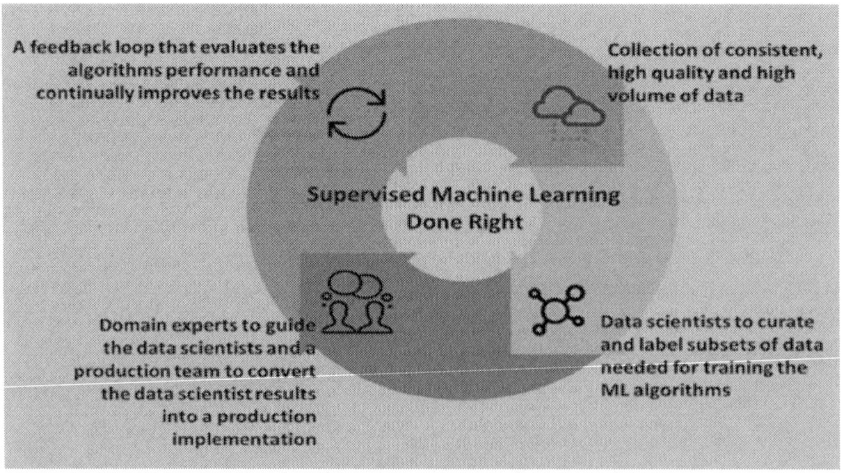

Fig. 2: is a schematic block diagram illustrating one embodiment of a data intelligence module.

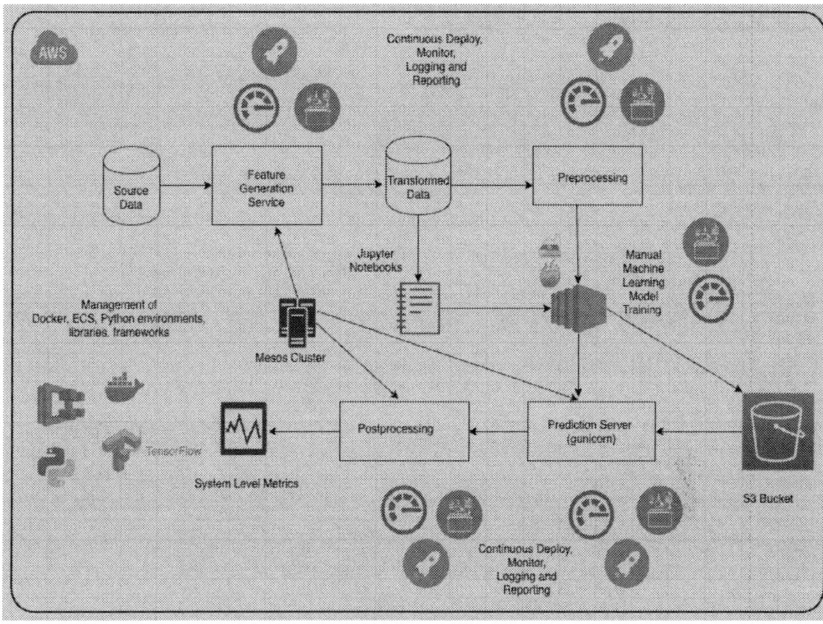

Fig. 3: is a schematic block diagram illustrating one embodiment of a system for a machine learning factory.

DESCRIPTION

Huge Data Engineering Big Data Engineering is the capacity and information control innovations that influence an assortment of evenly coupled assets to accomplish an almost straight versatility in execution. New designing methods in the information layer have been driven by the developing noticeable quality of information types that can't be dealt with effectively in a conventional social model. The requirement for versatile access in organized and unstructured information has prompted programming based on name-esteem/key-esteem sets or columnar (enormous table), report situated, and diagram (counting triple-store) ideal models.

Non-Relational Model : Non-Relational Models alludes to coherent information models like record, diagram, key worth and others that are utilized to give more proficient stockpiling and admittance to no even informational collections.

- **NoSQL:** NoSQL (then again called "no SQL" or "not just SQL") alludes to information stores and interfaces that are not attached to severe social methodologies.

- **Huge Data Models:** Big Data Models alludes to coherent information models (social and non-social) and preparing/ calculation models (bunch, streaming, and conditional) for the capacity and control of information across on a level plane scaled assets.
- **Blueprint on-read huge information:** are regularly put away in a crude structure dependent on its creation, with the composition, required for getting sorted out (and frequently purging) the information, is found and changed as the information are questioned. This is basic since all together for some examination to run effectively the information should be organized to help the particular calculations or preparing structures included.
- **Huge Data Analytics:** Big Data Analytics is quickly advancing both as far as usefulness and the hidden programming model. Such scientific capacities support the coordination of results inferred in equal across disseminated bits of at least one information sources.
- **Huge Data Paradigm:** The huge information worldview comprises of the dissemination of information frameworks across on a level plane coupled free assets to accomplish the adaptability required for the productive preparing of broad informational collections. With the new Big Data Paradigm, insightful capacities can be executed against the whole informational index or even progressively on a ceaseless stream of information.

Examination might even incorporate numerous information sources from various associations. For instance, think about the inquiry "What is the connection be tween's creepy crawly borne infections, temperature, precipitation, and changes in foliage". To respond to this inquiry an examination would have to incorporate information about rate and area of infections, climate information, and ethereal photography.

While we positively anticipate a proceeded with advancement in the strategies to accomplish proficient adaptability across assets, this changes in outlook (in relationship to the earlier change in the reproduction local area) is a one-time event; essentially until another change in outlook happens past this "publicly supporting" of preparing or information framework across various on a level plane coupled assets. The Big Data worldview has different ramifications from these specialized advancements.

The progressions are in the coherent information stockpiling, yet in the equal circulation of information and code in the actual record framework and direct inquiries against this stockpiling. The change in reasoning causes changes in the conventional information lifecycle. One portrayal of the start to finish information lifecycle arranges the means as assortment, readiness, investigation and activity. Diverse enormous information use cases can be described as far as the informational collection attributes very still or moving, and as far as the time window for the start to finish information lifecycle. Informational collection attributes change the information lifecycle measures in an unexpected way, for instance in the mark of a lifecycle at which the information are put in tireless capacity.

In a customary social model, the information is put away after groundwork (for instance after the concentrate change burden and purging cycles). In a high speed use case, the information is ready and investigated for cautioning, and really at that time is the information (or totals of the information) given a diligent stockpiling. In a volume use case the information is frequently put away in the crude state wherein it was created, preceding the utilization of the readiness cycles to purge and arrange the information. The outcome of perseverance of information in its crude state is that a blueprint or model for the information are possibly applied when the information are recovered, known as diagram on read.

A third outcome of huge information designing is frequently alluded to as "moving the preparing to the information, not the information to the handling". The ramifications is that the information are too broad to even think about being questioned and sent into one more asset for examination, so the investigation program is rather dispersed to the information holding assets; with just the outcomes being accumulated on an alternate asset. Since I/O transfer speed is every now and again the restricted asset in moving information, another methodology is insert question/channel programs inside the actual stockpiling medium.

At its heart, Big Data alludes to the expansion of information storehouses and preparing across evenly scaled assets, much similarly the figure escalated re-enactment local area accepted hugely equal handling twenty years prior. Previously, exemplary equal registering applications used a rich arrangement of communi.

Conversely, today, with informational indexes developing into the Petabyte and Exabyte scales, disseminated preparing systems offering examples, for example, map-lessen, offer a solid significant level, economically practical process model dependent on ware figuring assets, dynamic asset planning, and synchronization strategies Definition: Big Data is an information set(s) with qualities (for example volume, speed, assortment, changeability, veracity, and so forth) that for a specific issue area at a given point in time can't be productively handled utilizing current/existing/set up/customary advances and procedures to separate worth.

The above definition recognizes Big Data from business insight (BI) and customary value-based preparing while at the same time implying an expansive range of utilizations that incorporates them. A definitive objective of handling Big Data is to determine separated worth that can be trusted (on the grounds that the basic information can be trusted). This is done through the use of cutting edge examination against the total corpus of information paying little mind to scale. Parsing this objective aide outline the worth conversation for Big Data use cases.

- Any size of activities and information: Utilizing the whole corpus of applicable data, as opposed to simply tests or subsets. It's additionally about binding together all choice helps time-skylines (past, present, and future) through measurably determined bits of knowledge into profound informational indexes in that load of measurements.
- Reliable information: Deriving legitimate bits of knowledge either from a solitary form of-truth solidification and purifying of profound information, or from factual models that filter piles of "grimy" information to discover the needles of substantial understanding.
- Progressed investigation: Faster bits of knowledge through an assortment of logical and mining procedures from information designs, for example, "long tail" examinations, miniature divisions, and others, that are not doable in case you're compelled to more modest volumes, more slow speeds, smaller assortments, and dubious veracities.

A troublesome inquiry is what makes "Enormous Data" huge, or how huge does an informational index must be to be called large information? The appropriate response is an unacceptable "it depends". Some portion of this issue is that "Huge" is a general term and with the developing thickness of computational and capacity abilities (for example more force in more modest more productive structure factors) what is viewed as large today will probably not be viewed as large tomorrow. Information are thought of "large" if the utilization of the new adaptable structures gives further developed business productivity over other customary designs. As such, the usefulness can't be accomplished in something like a conventional social data set stage. Large information basically centers around oneself referring to perspective that information are huge on the grounds that it requires adaptable frameworks to deal with it, and structures with better scaling have come about as a result of the need to deal with enormous information

Parts of the current exposure might be epitomized as a device, framework, strategy, or PC program item. In like manner, parts of the current revelation might appear as a totally equipment epitome, a completely programming exemplification (counting firmware, occupant programming, miniature code, and so on) or an encapsulation joining programming and equipment angles that may all by and large be alluded to thus as a "circuit," "module" or "framework." Furthermore, parts of the current divulgence might appear as a PC program item epitomized in at least one PC decipherable capacity media having PC clear program code typified subsequently.

A significant number of the useful units depicted in this detail have been marked as modules, to all the more especially underscore their execution freedom. For instance, a module might be executed as an equipment circuit involving custom VLSI circuits or door exhibits, off-the-rack semiconductors, for example, rationale chips, semiconductors, or other discrete parts. A module may likewise be carried out in programmable equipment gadgets, for example, field programmable door clusters, programmable exhibit rationale, programmable rationale gadgets or something like that.

Modules may likewise be carried out in programming for execution by different kinds of processors. A distinguished module of executable code may, for example, include at least one physical or consistent squares of PC guidelines which may, for example, be coordinated as an article, methodology, or capacity. In any case, the executables of a recognized module need not be truly found together, however may involve dissimilar guidelines put away in various areas which, when consolidated legitimately, contain the module and accomplish the expressed reason for the module.

To be sure, a module of executable code might be a solitary guidance, or numerous directions, and may even be conveyed more than a few diverse code sections, among various projects, and across a few memory gadgets. Essentially, functional information might be recognized and represented thus inside modules, and might be typified in any reasonable structure and coordinated inside any appropriate kind of information structure.

The functional information might be gathered as a solitary informational collection, or might be appropriated over various areas including over various capacity gadgets, and may exist, unquestionably somewhat, simply as electronic signs on a framework or organization. Where a module or segments of a module are carried out in programming, the product segments are put away on at least one PC decipherable capacity media.

Any mix of at least one PC clear stockpiling media might be used. A PC lucid capacity medium might be, for instance, however not restricted to, an electronic, attractive, optical, electromagnetic, infrared, or semiconductor framework, mechanical assembly, or gadget, or any reasonable blend of the previous.

PC program code for completing activities for parts of the current exposure might be written in any mix of at least one programming dialects, including an item arranged programming language like Java, Python, C++ or something like that and traditional procedural programming dialects, for example, the "C" programming language or comparative programming dialects. The program code might execute completely on the client's PC, somewhat on the client's PC, as an independent programming bundle, mostly on the client's PC and incompletely on a far off PC or altogether on the far off PC or worker. In the last situation, the distant PC might be associated with the client's PC through an organization, including a neighborhood (LAN) or a wide region organization (WAN), or the association might be made to an outer PC (for instance, through the Internet utilizing an Internet Service Provider).

Reference all through this detail to "one encapsulation," "an epitome," or comparable language implies that a specific element, design, or trademark portrayed regarding the exemplification is remembered for somewhere around one encapsulation of the current exposure. Along these lines, appearances of the expressions "in one epitome," "in an encapsulation," and comparative language all through this determination may, however don't really, all allude to a similar exemplification, yet signify "at least one yet not all exemplifications" except if explicitly indicated something else.

The expressions "counting," "containing," "having," and varieties thereof signify "counting yet not restricted to" except if explicitly determined something else. A counted posting of things doesn't infer that any or the entirety of the things are totally unrelated or potentially comprehensive together, except if explicitly indicated something else. The expressions "a," "an," and "the" likewise allude to "at least one" except if explicitly indicated something else.

Moreover, the portrayed components, designs, or attributes of the exposure might be consolidated in any reasonable way in at least one encapsulations. In the accompanying depiction, various explicit subtleties are given, like instances of programming, programming modules, client determinations, network exchanges, data set questions, information base constructions, equipment modules, equipment circuits, equipment chips, and so forth, to give a careful comprehension of exemplifications of the divulgence. Be that as it may, the divulgence might be drilled without at least one of the particular subtleties, or with different techniques, parts, materials, etc. In different examples, notable constructions, materials, or activities are not shown or depicted exhaustively to try not to cloud parts of the revelation.

Parts of the current revelation are depicted underneath concerning schematic flowchart charts and additionally schematic square outlines of techniques, devices, frameworks, and PC program items as indicated by epitomes of the divulgence. It will be perceived that each square of the schematic flowchart charts or potentially schematic square outlines, and mixes of squares in the schematic flowchart graphs as well as schematic square graphs, can be carried out by PC program guidelines.

These PC program directions might be given to a processor of a universally useful PC, particular reason PC, or other programmable information handling device to deliver a machine, with the end goal that the guidelines, which execute through the processor of the PC or other programmable information preparing mechanical assembly, make implies for carrying out the capacities/acts indicated in the schematic flowchart outlines and additionally schematic square charts square or squares.

These PC program guidelines may likewise be put away in a PC decipherable capacity medium that can coordinate a PC, other programmable information preparing mechanical assembly, or different gadgets to work in a specific way, with the end goal that the directions put away in the PC meaningful capacity medium produce an article of production including guidelines which carry out the capacity/act indicated in the schematic flowchart outlines and additionally schematic square graphs square or squares.

The PC program directions may likewise be stacked onto a PC, other programmable information handling device, or different gadgets to make a progression of functional advances be performed on the PC, other programmable device or different gadgets to deliver a PC carried out interaction to such an extent that the guidelines which execute on the PC or other programmable contraption give cycles to carrying out the capacities/acts determined in the flowchart and additionally block chart square or squares.

The schematic flowchart charts or potentially schematic square outlines in the Figures show the design, usefulness, and activity of potential executions of contraptions, frameworks, strategies and PC program items as per different encapsulations of the current divulgence. In such manner, each square in the schematic flowchart outlines as well as schematic square charts might address a module, section, or piece of code, which includes at least one executable directions for carrying out the predefined consistent function(s).

It ought to likewise be noticed that, in some elective executions, the capacities noted in the square might happen out of the request noted in the figures. For instance, two squares displayed in progression may, truth be told, be executed significantly simultaneously, or the squares may in some cases be executed in the opposite request, contingent on the usefulness in question. Different advances and techniques might be considered that are identical in capacity, rationale, or impact to at least one squares, or partitions thereof, of the represented figures.

Albeit different bolt types and line types might be utilized in the flowchart and additionally block graphs, they are perceived not to restrict the extent of the relating encapsulations. To be sure, a few bolts or different connectors might be utilized to show just the coherent progression of the portrayed encapsulation. For example, a bolt might show a pausing or observing time of vague length between identified stages of the portrayed encapsulation. It will likewise be noticed that each square of the square outlines as well as flowchart graphs, and blends of squares in the square charts or potentially flowchart outlines, can be carried out by particular reason equipment based frameworks that play out the predetermined capacities or acts, or mixes of specific reason equipment and PC directions. The portrayal of components in each figure might allude to components of continuing figures. Like numbers allude to like components in all figures, including substitute encapsulations of like components.

Hadoop: is a free Java-based programming system upholds the preparing of huge arrangements of information in a dispersed figuring climate. It is a piece of the Apache project supported by the Apache Software Foundation. Hadoop group utilizes a Master/Slave structure. Utilizing Hadoop, huge informational indexes can be handled across a group of workers and applications can be run on frameworks with a great many hubs including a large number of terabytes.

Disseminated record framework in Hadoop helps in quick information move rates and permits the framework to proceed with its ordinary activity even on account of some hub disappointments. This methodology brings down the danger of a whole framework disappointment, even on account of a critical number of hub disappointments. Hadoop empowers a figuring arrangement that is versatile, savvy, and adaptable and issue open minded. Hadoop Framework is utilized by well known organizations like Google, Yahoo, Amazon and IBM and so on, to help their applications including gigantic measures of information. Hadoop has two primary sub undertakings – Map Reduce and Hadoop Distributed File System (HDFS).

Guide Reduce Hadoop Map: Reduce is a system used to compose applications that interaction a lot of information in equal on groups of item equipment assets in a dependable, issue open minded way. A Map Reduce work first partitions the information into individual lumps which are prepared by Map occupations in equal. The yields of the guides arranged by the structure are then contribution to the diminish errands. By and large, the information and the yield of the work are both put away in a record framework. Booking, Monitoring and once again executing bombed undertakings are taken consideration by the system.

Hadoop Distributed File System: (HDFS) HDFS is a record framework that traverses every one of the hubs in a Hadoop bunch for information stockpiling. It connects together record frameworks on nearby hubs to make it into one enormous document framework. HDFS further develops dependability by recreating information across different sources to conquer hub disappointments.

Benefits of Big Data Applications: The huge information application alludes to the huge scope appropriated applications which normally work with huge informational collections. Information investigation and examination transformed into a troublesome issue in numerous areas in the range of huge information. With huge and complex information, calculation becomes hard to be taken care of by the conventional information preparing applications which triggers the advancement of enormous information applications. Google's guide decrease structure and apache Hadoop are the deserter programming frameworks for enormous information applications, in which these applications creates an immense measure of moderate information. Assembling and Bioinformatics are the two significant spaces of enormous information applications.

There are many benefits of coordinating distributed computing to huge date. Huge information raises the requirement for different workers due to the enormous information and size it deals with, and it requests high speed and changeability. These various workers work in corresponding to give the large information high necessities. Distributed computing as of now utilizes numerous workers and permit asset designations. Therefore, it is an incredible fit to fabricate the large information on these cloud multi-workers and utilize the asset distribution accessibility given by the cloud conditions which would bring about a superior productivity for huge information examination.

Utilizing cloud framework as a capacity for huge information would work on the presentation for both. As cloud frameworks are principally founded on remote multi-workers which makes it practical to deal with enormous measures of information all the while. This element permits progressed investigation methods to empower the large information to manage gigantic measures of information. The combination between distributed computing and enormous information would bring about cost decrease. While huge information requires bunches of workers and volumes to help the gigantic measure of information. Distributed computing frameworks can function as the construction for these workers and volumes as opposed to making new ones for large information which additionally gives greater adaptability and versatility and dispenses with the gigantic ventures that would be contributed on the enormous information PCs and workers.

Likewise, utilizing distributed computing gives quicker provisioning to large information as provisioning workers in the cloud is so natural and doable. Along these lines, in view of the handling prerequisites of the huge information, the cloud climate utilized can be scaled as needs be. This quick provisioning is truly significant for enormous information as the worth of the information quickly decline after some time. By and large, distributed computing supplements enormous information and gives advantageous, on-request and imparted figuring climate to insignificant administration exertion and diminished overhead. It likewise makes the climate more vigorous, robotized and gives multi-occupancy. Notwithstanding the entirety of that, the joining between both, makes large information assets more controllable, observed and detailed. Additionally, this coordination empowers lessening the intricacy and working on the usefulness. These benefits make the cloud based methodologies are the right models for conveying huge information

Our development" BIG DATA AND CLOUD COMPUTING." is a Big information is presently perhaps the most basic arising technology. Huge Data are utilized as an idea that alludes to the failure of conventional information models to proficiently deal with the new informational indexes. The 4V's of enormous information – volume, speed, assortment and veracity makes the information the executives and investigation trying for the conventional information distribution centers. Consider huge information and investigation together. Huge information is the term used to depict the new blast of various sorts of information from different sources. Examination is tied in with analyzing information to infer fascinating and applicable patterns and examples, which can be utilized to educate choices, enhance measures, and even drive new plans of action. Distributed computing is by all accounts an ideal vehicle for facilitating enormous information jobs. In any case, chipping away at large information in the cloud brings its own test of accommodating two conflicting plan standards. Distributed computing depends on the ideas of solidification and asset pooling, yet large information frameworks (like Hadoop) are based on the common nothing rule, where every hub is free and independent. The coordinating enormous information with distributed computing advances, organizations and training establishments can have a superior heading to what's to come. The capacity to store a lot of information in various structures and interaction everything at exceptionally huge rates will bring about information that can direct organizations and schooling foundations in growing quick. By and by, there is a huge concern in regards to protection and security issues when moving to the cloud which is the primary driver with respect to why organizations and instructive establishments won't move to the cloud. This paper presents the qualities, patterns and difficulties of huge information. Likewise, it examines the advantages and the dangers that might emerge from the combination between large information and distributed computing.

OUTCOME

Our creation" IML-Cloud Data Performance" is a framework, strategies, and PC program items are introduced for performing cloud information execution and examination utilizing AI. The created innovation alos solo and managed learning module is designed to collect an unstructured informational index into numerous renditions of a coordinated informational index. The IML-cloud information execution is an administered learning module is designed to create more the one AI gatherings dependent on every variant of numerous renditions of a coordinated informational index and to figure out which AI troupe shows a most elevated prescient exhibition. The IML-cloud information execution is a technique for performing cloud information execution and examination utilizing AI and furthermore a strategy, incorporates extricating information from at least one cloud information sources. The designed innovation additionally incorporates stacking cloud information into an unstructured informational index having an unstructured organization and a strategy in certain incorporates collecting an unstructured informational collection into a coordinated informational collection having an organized arrangement and furthermore a technique incorporates creating at least one learned capacities dependent on a coordinated informational collection.

1. According to claim1# the creation is to a framework, techniques, and PC program items are introduced for performing cloud information execution and examination utilizing AI.
2. According to claim1,2# the creation is to an unaided and managed learning module is arranged to gather an unstructured informational collection into numerous variants of a coordinated informational collection.
3. According to claim1,2,3# the creation is to a cloud information execution is a regulated learning module is arranged to produce more the one AI gatherings dependent on every adaptation of different renditions of a coordinated informational collection and to figure out which AI troupe shows a most elevated prescient exhibition.
4. According to claim1,2,3,4# the innovation is to a will be a technique for performing cloud information execution and examination utilizing AI and furthermore a strategy, incorporates removing information from at least one cloud information sources?
5. According to claim1,2,3,5# the innovation is to a stacking cloud information into an unstructured informational collection having an unstructured organization and a technique in certain incorporates gathering an unstructured informational collection into a coordinated informational index having an organized configuration and furthermore a strategy incorporates creating at least one learned capacities dependent on a coordinated informational index.

REFERENCES

[1] Abhigna, B.S., Chakraborty, S., Kumar, V., Medha: A collaborative E-concept framework, Proceedings of the 2018 International Conference On Communication, Computing and Internet of Things, IC3IoT 2018, 2019, pp. 117–122, 8668127

[2] Abidin, S., Vadi, V.K., Rana, A., On Confidentiality, Integrity, Authenticity, and Freshness (CIAF) in WSN, Advances in Computer, Communication and Computational Sciences, 2020, pp 87-97

[3] Ahay, A., Choudhury, T., Kumar, V., Hierarchical Evolution of Digital Games, Proceedings of 2019 International Conference on Computational Intelligence and Knowledge Economy, ICCIKE 2019, 2019, pp. 381–386, 9004378

[4] Area: Communication, Frequency modulation, Networks, wireless communication

[5] Area: Communication, Frequency modulation, Networks, wireless communication

[6] Area: Communication, Frequency modulation, Networks, wireless communication

[7] Area: Communication, Frequency modulation, Networks, wireless communication

[8] Arora, M., Kumar, V., Dave, M., Hybrid cloud infrastructure for load balancing and scaling with vm path optimization using ant colony optimization, Journal of Advanced Research in Dynamical and Control Systems, 2020, 12(5 Special Issue), pp. 660–674

[9] Arora, M., Kumar, V., Dave, M., Task scheduling in cloud infrastructure using optimization technique genetic algorithm, Proceedings of the World Conference on Smart Trends in Systems, Security and Sustainability, WS4 2020, 2020, pp. 788–793, 9210303

[10] Bansal, R., Gupta, A., Singh, R., Nassa, V.K., Role and impact of digital technologies in E-learning amidst COVID-19 pandemic, Proceedings - 2021 4th International Conference on Computational Intelligence and Communication Technologies, CCICT 2021, 2021, pp. 194–202

[11] Bansal, R., Obaid, A.J., Gupta, A., Singh, R., Pramanik, S., Impact of Big Data on Digital Transformation in 5G Era, Journal of Physics: Conference Seriesthis link is disabled, 2021, 1963(1), 012170

[12] Chauhan, S.S., Khare, A.K., Kinematic analysis of the ABB IRB 1520 industrial robot using roboanalyzer software, Evergreenthis link is disabled, 2020, 7(4), pp. 510–518

[13] Chauhan, S.S., Khare, A.K., Method of Obtaining Planar State of Stress Using Mohr's Circle—Some Typical Cases, Lecture Notes in Mechanical Engineering, 2022, pp. 33–38

[14] Goel, S., Tripathi, A., Choudhury, T., Kumar, V. Parkinson's Disease Detection using Soft Computing Technique, Proceedings of the 2019 8th International Conference on System Modeling and Advancement in Research Trends, SMART 2019, 2020, pp. 3–6, 9117384

[15] Govindarajan, V., Sivakumar, R., Patil, P.P., Kannan, M., Ramesh, B., Effect of Tungsten Carbide Addition on the Microstructure and Mechanical Behavior of Titanium Matrix Developed by Powder Metallurgy Route, Advances in Materials Science and Engineeringthis link is disabled, 2022, 2022, 2266951

[16] Gupta, A., Singh, R., Nassa, V.K., Sharma, P., Koti, K., Investigating Application and Challenges of Big Data Analytics with Clustering, 2021 International Conference on Advancements in Electrical, Electronics, Communication, Computing and Automation, ICAECA 2021, 2021

[17] http://ijcsit.com/docs/Volume%206/vol6issue03/ijcsit2015060323.pdf

[18] https://acme.able.cs.cmu.edu/pubs/uploads/pdf/IoTBD_2016_10.pdf

[19] https://www.ijser.org/researchpaper/Big-Data-Analytics-inCloud-Computing-Environment.pdf

[20] https://www.intel.in/content/www/in/en/now/edge-to-cloud/cloud.html?tracksrc=2021Q3_e2c_sh_cloud_create_tx&cid=sem&source=sa360&campid=2021_q3_dpg_in_dpgeng_dpgeng2

[21] https://www.researchgate.net/publication/325011725_Review_Paper_on_Big_Data_Analytics_in_Cloud_Computing

[22] https://www.tandfonline.com/doi/full/10.1080/17538947.2016.1239771

[23] Jain, P., Vijay, S., & Gupta, S. C. (2009). ATM Network QoS Parameters and Performance Analysis using OPNET Simulator. International Journal of Recent Trends in Engineering, 2(4), 228.

[24] Jain, P., Vijay, S., & Gupta, S. C. Performance Analysis & QoS Guarantee in ATM Networks. Global Journal of computer Science and Technology, 131-136.

[25] Jain, S., Choudhury, T., Kumar, V., Kumar, P., Detecting malware and analysing using sandbox evasion, Proceedings of the 2018 International Conference On Communication, Computing and Internet of Things, IC3IoT 2018, 2019, pp. 111–116, 8668149

[26] Kala, S., Nandan, H., Sharma, P., Shadow and weak gravitational lensing of a rotating regular black hole in a non-minimally coupled Einstein-Yang-Mills theory in the presence of plasma, European Physical Journal Plusthis link is disabled, 2022, 137(4), 457

[27] Kandpal, H., Kumar, A., Malik, A., Morphometric analysis and prioritization of sub-watersheds in Bino watershed, Uttarakhand: A remote sensing and GIS perspective, Indian Journal of Ecology, 2018, 45(3), pp. 475–478

[28] Kannan, M., Singh, S., Prasad, R.R., Synthetic Methods to Obtain Calcia-Stabilized Zirconia Powders—A Review, Lecture Notes in Mechanical Engineeringthis link is disabled, 2021, pp. 405–416

[29] Kukreti, A., Painolli, A.K., Rana, N., Where do we stand: factors affecting sustainable development, Procedia Environmental Science, Engineering and Management, Volume 8, Issue 2, 2021, Pages 583-589

[30] Kulkarni, S & Chima, P 2021 'Challenges faced by UK university students due to thcoronavirus crisis in the Higher Education' Preprints.org. https://doi.org/10.20944/preprints202102.0192.v1

[31] Kulkarni, S & Sodre, JR 2017, An Assessment of Lightweight Vehicles to Reduce Greenhouse Gas Emissions with Focus on Magnesium. in Thoughts and Refections on the use of Magnesium. Birmingham City University, pp. 84-91.

[32] Kulkarni, S, Chapman, C & Shah, H 2016 'Computational Fluid Dynamics (CFD) Mesh Independency Study of A Straight Blade Horizontal Axis Tidal Turbine' Preprints.org. https://doi.org/10.20944/preprints201608.0008.v1

[33] Kulkarni, S, Chapman, C, Shah, H & Edwards, D 2018, 'A computational design method for bio-mimicked horizontal axis tidal turbines', International Journal of Building Pathology and Adaptation, vol. 36, no. 2, pp. 188-209. https://doi.org/10.1108/IJBPA-06-2017-0029

[34] Kulkarni, S, Chapman, C, Shah, H, Parn, E & Edwards, DJ 2018, 'Designing an efficient tidal turbine blade through bio-mimicry: A systematic review', Journal of Engineering, Design and Technology, vol. 16, no. 1, pp. 101-124. https://doi.org/10.1108/JEDT-08-2017-0077

[35] Kulkarni, S, Chapman, C, Shah, H, Parn, E & Edwards, DJ 2018, 'Design Study of Horizontal Axis Tidal Turbine Blade', Mindanao Journal of Science and Technology. https://www.open-access.bcu.ac.uk/10089/

[36] Kulkarni, S, Edwards, DJ, Chapman, C, Hosseini, MR & Owusu-Manu, D-G 2019, 'A preliminary mechanical design evaluation of the Wikispeed car: for light-weighting implications', Journal of Engineering, Design and Technology, vol. 17, no. 1, pp. 230-249. https://doi.org/10.1108/JEDT-09-2018-0154

[37] Kulkarni, S, Edwards, DJ, Parn, E, Chapman, C, Aigbavboa, CO & Cornish, R 2018, 'Evaluation Of Vehicle Lightweighting To Reduce Greenhouse Gas Emissions With Focus On Magnesium Substitution', Journal of Engineering, Design and Technology, vol. 16, no. 6, pp. 869-888. https://doi.org/10.1108/JEDT-03-2018-0042

[38] Kulkarni, SS, Wang, L, Golsby, N & Lander, M 2021, 'Fluid-structure interaction based optimisation in tidal turbines: A perspective review', Journal of Ocean Engineering and Science. https://doi.org/10.1016/j.joes.2021.09.017

[39] Malik, A., Kumar, A., Kandpal, H., Morphometric analysis and prioritization of sub-watersheds in a hilly watershed using weighted sum approach, Arabian Journal of Geosciencesthis link is disabled, 2019, 12(4), 118

[40] Mohanakurup, V.K., Baghela, V.S., Kumar, S., Srivastava, P.K., Doohan, N.V., Soni, M., Awa, H., 5G Cognitive Radio Networks Using Reliable Hybrid Deep Learning Based on Spectrum Sensing, Green Internet of Things in 5G and Beyond, Hindawi, Volume 2022

[41] Mohta, M., Soni, N., Choudhury, T., Kumar, V., Market Analysis on Li-Fi Technology, Proceedings of the 2019 8th International Conference on System Modeling and Advancement in Research Trends, SMART 2019, 2020, pp. 139–143, 9117427

[42] Nagwan, H., Rana, N., Kumar, A., Tewari, P.C., Performance and emission characteristics of diesel engine using Karanja oil methyl ester and Eucalyptus oil fuel blends, Procedia Environmental Science, Engineering and Management, Volume 6, Issue 4, 2019, Pages 553-563

[43] Painoli, A.K., Bansal, R., Singh, R., Kukreti, A., Impact of digital marketing on the buying behavior of youth with special reference to Uttarakhand state, Big Data Analytics for Improved Accuracy, Efficiency, and Decision Making in Digital Marketing, 2021, pp. 162–182

[44] Parag, J. (2011). Hybrid mixed handover for call blocking probability analysis in wireless ATM network. Journal of Engineering and Computer Innovations, 2(4), 68-77.

[45] Sharma, C., Singh, M.P., Chaudhary, S., Cost Analysis of Solar-Wind Based Hybrid Renewable Energy System, Lecture Notes in Electrical Engineeringthis link is disabled, 2021, 700, pp. 2733–2746

[46] Sharma, S., Chandra, U., & Jain, P. (2017). A literature survey on automation of test data generation for branch coverage testing using genetic algorithm. International Journal of Computational Intelligence Research, 13(6), 1521-1531.

[47] Varshney, R., Jain, P., & Vijay, S. (2018, September). Massive MIMO Systems In Wireless Communication. In 2018 2nd International Conference on Micro-Electronics and Telecommunication Engineering (ICMETE) (pp. 39-44). IEEE.

CHAPTER - 07

Passive Direct Methanol Fuel Cell using Machine Learning

Dr. Sanjay B. Waykar

Mahatma Gandhi Mission's College of Engineering and Technology, India.
Email:sbwaykar@gmail.com

Dr. Anil W. Kale

Mahatma Gandhi Mission's College of Engineering and Technology, India.
Email: anil5474@gmail.com

Dr. Nandkishor P Karlekar

MGMCET Kamothe Navimumbai, India. Email: npkarlekar.mgmcet@gmail.com

Dr. Vaibhav Narawade

Ramrao Adik Institute of Technology, India. Email:vnarawade@gmail.com

ABSTRACT

Our Invention "Passive Direct Methanol Fuel Cell using Machine Learning " is a creation connects with performance and fundamental perspectives concerning the Passive Direct Methanol Fuel Cell innovation that are mainly focusing on Membrane Electrode Assembly effective areas and perimeters. Hence it is required to analyze the best performance and to find out the highest power density of cell using various shapes by altering gasket openings of dynamic regions with equal areas and equal perimeters. Comparison of the shapes such as square, rectangular, rhombus, and circular active area forms and by using Nickel as current collector material is selected for evaluation. In the equal areas with gasket opening area of 1963.5 (+/-0.2) mm^2, among circular, rhombus, rectangular and square shapes chosen for investigation on the performance of cell at 5 molar methanol concentration, square shape opening consisting of the *perimeter of 177.2 mm has developed highest power density of 6.344 mWcm^{-2} with a corresponding maximum current density of 65.2 mAcm^{-2}. In the equal perimeters of 157.1 (+/- 0.2) mm of gasket openings, among circular, rhombus, rectangular and square shapes chosen for investigation on the performance of cell at 5 molar methanol concentration, Rhombus shape opening consisting of area 1400 mm^2 has developed the highest power density of 7.714 mWcm^{-2} with a corresponding maximum current density of 85.3 mAcm^{-2}. In the large active areas, methanol consumption is slow because the current density is lower when compared to the smaller active areas. As a result, the surface-specific concentration on the ion-conducting membrane is being more, making a higher crossover of methanol over bigger active zones. The experimental investigation demonstrated that the size of the dynamic zone significantly affects the performance of the cell and its performance can be improved with the expansion in dynamic region size. In the vast dynamic regions, the current density is lower compared to the smaller dynamic zones where the methanol is expended gradually. So, the local concentration on the cell membrane is more and brings further fuel cross-over. The exothermic response that is involved on penetrated fuel and oxidant on cathode impetus generates more heat, which prompts an elevated working temperature. This elevated working temperature brings about superior electrochemical kinetics of fuel oxidation and a decrease in oxygen response, leading to superior cell performance. The internal obstruction decreases with expanding the size of the dynamic zone. This reality is good with the accomplished outcome. Even though the cell performance can be enhanced by expanding the size of the dynamic zone, with the research outcome as on date, it is revealed that tiny dynamic zones have its focal points, for example, more slow charge transfer obstruction, lesser flooding, even distribution of methanol dispersion, increasing of effective current gathering, minor expenditure on fabrication, and further, it is small in size and weight.*

It is also learned from the experiment that cell performance depends on the combined effect of (i) gasket opening shape, opening area, and its perimeter; (ii) the free open ratio and opening perimeter on the current collector plates. Comparison study referring to power and power density on dynamic section shows the influence of shapes and sizes of the geometry of effective MEA regions.

The results obtained in the study show that the fuel cell performance is a function of membrane-active area size and shapes; current collectors opening area size and shapes. The experimental results indicate that with equal areas, the cell performance is best using square shape compared to circle, rhombus, and rectangle. However, in the equal perimeters, the cell performance is best using square shape compared to circle, rhombus, and rectangle.

BACKGROUND

Passive Direct Methanol Fuel Cell (PDMFC) is appropriate for the most range which includes portable and stationary electric & electronics applications as it is a better promising and emerging one, among the power sources in renewable energies. The performance of PDMFC hydrogen ion or proton-conducting electrolyte is dependent totally on polymer electrolyte membrane fuel cell technology.

Direct Methanol Fuel Cell which is portrayed as a gadget that changes over substance vitality into electrical vitality is a promising force source that relies upon the layer and the electro-impetus. The serious issues of the PDMFC are the methanol traverse which discourages cathode performance and declines the fuel effectiveness; this issue can be focused through the advancement of new impetus.

PRIOR ART SEARCH

In PDMFC, a proton-leading polymer membrane covered with impetus on the two sides is applied. This unit is known as a membrane electrode assembly. Polymer electrolyte membrane fuel cells use a strong particle conductive polymer as an electrolyte. Not only the proton conductivity but has it also made physical division of the anode and cathode chambers.

Electric protection and low electrical impedance are the principal significant properties of the electrolyte. The most widely recognized membranes have polymeric perfluoro sulfonic functional derivates, which are fabricated with essential additives for conducting ions.

To analyze the best performance and to find out the highest power density of cell using various shapes by altering gasket openings of dynamic regions with equal areas and equal perimeters. Comparison of the shapes such as square, rectangular, rhombus, and circular active area forms and by using Nickel as current collector material is selected for evaluation.

LITERATURE REVIEW

Vasco S. Silva1, et al. analyzed on characterization, experimentation, and modeling basis. A significant investigation is carried out on essential characteristics of the PDMFC system that focuses primarily on the membrane. This paper focuses chiefly on the PEM performance with an outline of research advancement. Detailed inputs are committed to researching related aspects over the preparation of membrane, its characterization, testing including modeling.

R. Govindarasu in their publication on "Recent evolutions in modeling of direct methanol fuel cell", stated that Direct Methanol Fuel Cells produce one of the better potential substitutes to the substitution of non-renewable energy sources. Anyway, whatever the efficient plan of the cell might be, the cell can't achieve success without legitimate modeling and simulation. The significance and the requirement for the modeling of a DMFC are talked about in detail. The modeling of the key segments, for example, the dispersion layer, MEA, stream dissemination, impetus thickness which add to the performance of the fuel cell are talked about.

Ritesh Pathak presents an analysis of direct methanol fuel cell (DMFC-work and DMFC-froth) frameworks interfaced with DC-DC converters. The yield voltage attributes of six frameworks are thought about. Exploratory outcomes show that DMFC frameworks present variable, nonstraight impedance as a capacity recurrence. It is appeared, through test results, that DMFC frameworks don't display unsurprising yield as the capacity of obligation cycle alone. Therefore, yield voltage control through heartbeat width regulation (PWM) procedures can't be utilized for DMFC frameworks. The inspiration for this analysis is a primer examination concerning the utilization of DMFCs as a force hotspot for a model Martian Uninhabited Aerial Vehicle (UAV).

Jing-Yi Chang, Yean-Der Kuan, and Shi-Min Lee carried out an Experimental Investigation of a Direct Methanol Fuel Cell with Hilbert Fractal Current Collectors. Fractal Current Collectors are fabricated with current collectors of various free opening perimeters and opening ratios. They had experimented on the investigation into the performance of PDMFC as a function of the opening perimeter and on the free open ratio of conducting bipolar plates. But these current collectors could not show better performance.

R. Hashemi & S. Yousefi& M. Faraji in their experimental study of the effect of the active area on the performance of the cell, have observed the methanol consumption over various active areas. Over the huge dynamic zones, as current density is lesser when compared to the lesser dynamic zones, fuel is absorbed gradually. Hence, the local fuel concentration over the polymer membrane is more, resulting in further fuel crossover.

OBJECTIVES

To analyze the best performance and to find out the highest power density of cell using various shapes by altering gasket openings of dynamic regions with equal areas and equal perimeters. Comparison of the shapes such as square, rectangular, rhombus, and circular active area forms and by using Nickel as current collector material is selected for evaluation.

- The objective of the invention is to an Experimental Investigation on Performance of Passive Direct Methanol Fuel Cell using dynamic active regions of Membrane Electrode Assembly with Nickel Current Collectors concerning effective equal areas and equal perimeters by masking membrane with varying shapes and sizes of gasket openings such as square, circle, rhombus, and rectangle.
- The other objective of the invention is to know the performance of the cell with the opening shape of MEA with equal areas and equal perimeters in terms of power densities and current densities.
- The other objective of the invention is the Comparison of gasket opening perimeter and area of effective Opening Shapes with equal areas and equal perimeters.
- The other objective of the invention is the Influence of gasket opening area to perimeter ratio (A/P) on effective gasket opening Shapes with equal areas and equal perimeters.
- The other objective of the invention is the Influence of the Current collector's opening perimeter and opening area on gasket effective Opening Shapes with equal areas and equal perimeters.
- The other objective of the invention is the Influence of the Current collector's opening area to opening perimeter ratio with equal areas and equal perimeters.
- The other objective of the invention is the Influence of the Current collector's opening ratio with equal areas and equal perimeters.
- The other objective of the invention is the Comparison of Power and Power Density on various Opening Shapes and sizes.

DIAGRAM

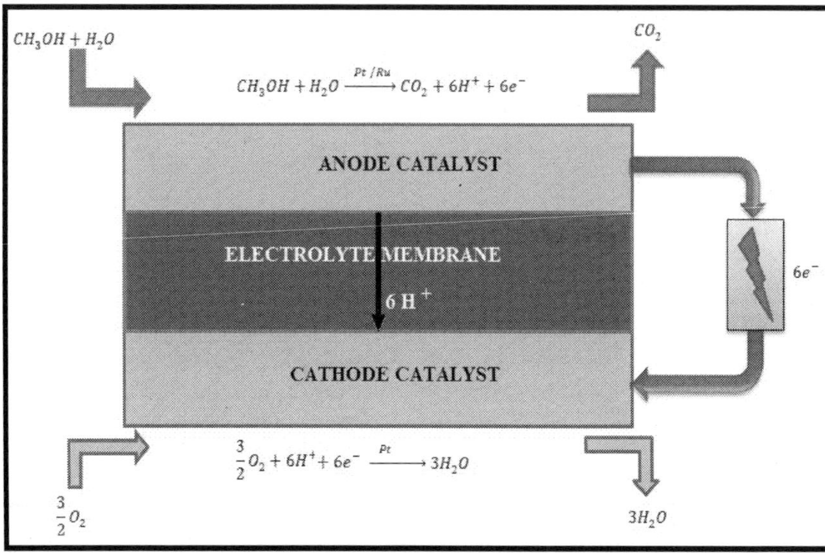

Fig. 1: Schematic presentation of the direct methanol fuel cell

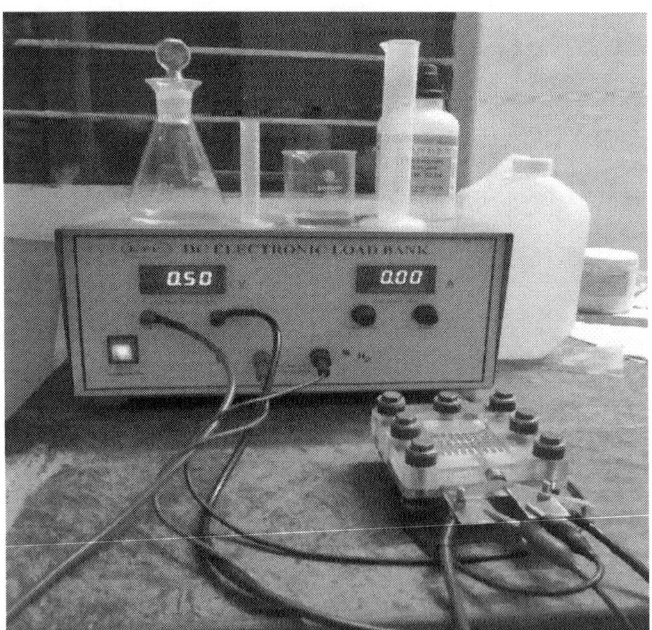

Fig. 2: Experimental Setup of PDMFC

Figure-3, Setup of the gasket over the anode-side current collector

Fig. 4: Setup of the gasket over MEA

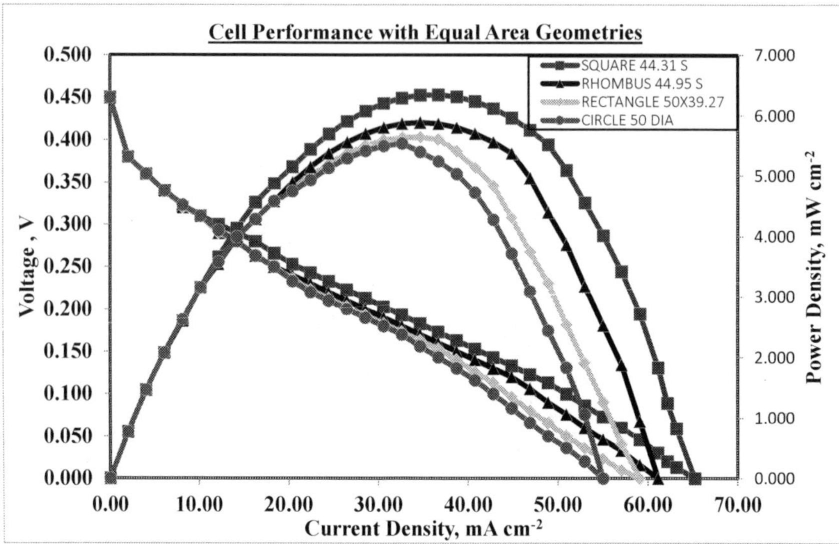

Fig. 5: Cell characteristics with various shapes of equal areas in gasket openings

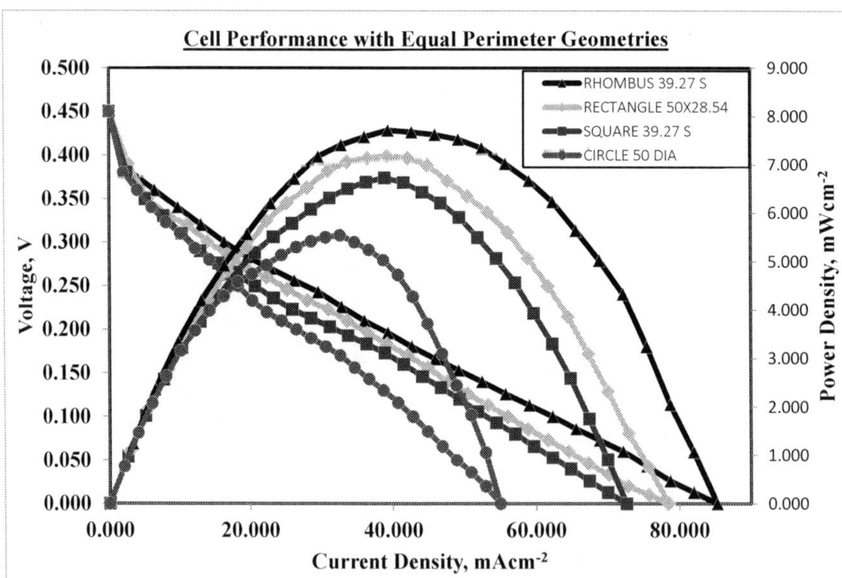

Fig. 6: Cell characteristics with various shapes of equal perimeters in gasket openings

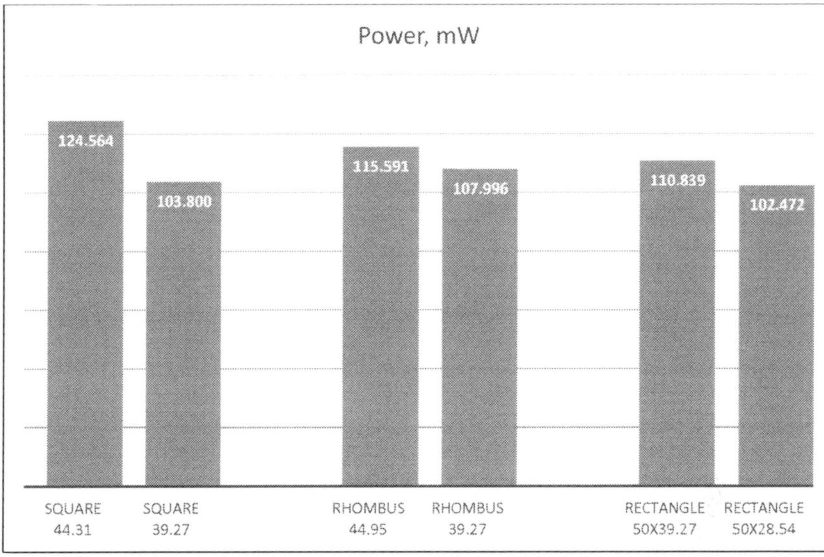

Fig. 7: Power produced by various shapes and sizes in gasket opening.

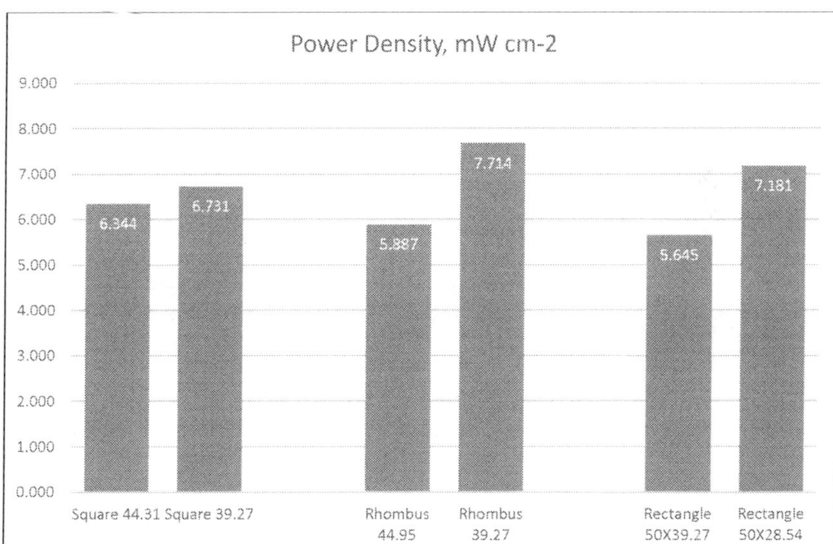

Fig. 8: Power density (bottom) produced by various shapes and sizes in gasket opening.

DESCRIPTION

The essential PDMFC is included by electrical terminals namely anode & cathode, with a strong electrolytic membrane in the middle. The typically used impetuses (catalyst) in PDMFC anode & cathode are Pt/Ru (~4 mg/cm^2) and Pt (~2 mg/cm^2) respectively. Relating to the electrolyte, PDMFC utilizes a proton exchange partition that isolates the anode and cathode; also permits the movement of hydrogen ions. Even though the essential thermodynamic qualities such as reversible oxidation potential and the methyl alcohol electro-oxidation potential, the mobilization of electrons to the anodic end for fuel conversion to carbon dioxide is also important. The comprised cell processes are shown as anodic end reaction equation (1), cathode end reaction as equation (2) and with overall reaction as equation (3). The schematic cell is shown in figure-1.

Anode End Reaction:

$$CH_3OH + H_2O \xrightarrow{Pt/Ru} CO_2 + 6H^+ + 6e^- \tag{1}$$

Catode End Reaction:

$$1.5\ O_2 + 6H^+ + 6e^- \xrightarrow{Pt} 3H_2O \tag{2}$$

Overall Fuel Cell Reaction:

$$CH_3OH + 1.5\ O_2 \rightarrow CO_2 + 2H_2O \tag{3}$$

EXPERIMENTAL SETUP

To investigate the performance of Passive Direct Methanol Fuel Cell using gaskets of different geometries having Equal Areas and Equal Perimeters with experimental evaluation, a single direct methanol fuel cell fixture is used with current collectors that fabricated using Ni material made with 2 mm thick plate. The current collector material of PDMFC is one of the vital elements that affect the cell performance characteristics. Besides its material for fabrication, design aspects, dimensions with shape factors also influence the cell performance. The essential characteristics of the current collector materials in PDMFC are:

- Better electrical conductivity.
- Optimum thermal conductivity.
- High tensile and flexural rigidity of the material.

- Ease of fabrication and machinability.
- Corrosion resistance.
- Longer durability.
- Less density.
- Ease of availability
- Low contact resistance and impedance.
- Even transport and distributing over an active region of reactants.

This experimental setup of the PDMFC is shown in Figure-2. Circular openings of 100 numbers, in 10 by 10 matrix patterns are made with a 3.8 mm diameter drill in the current collector. Openings percentage on current collectors is 45.3% in an effective 25 cm^2 area. A solid permeable membrane, Nafion-117 polymer electrolyte is used in MEA. The anode side is dopped with a catalyst around 4 mg cm^{-2} made up of Platinum-Ruthenium (in equal proportions) on Carbon cloth and on the cathode side, a catalyst of 2 mg cm^{-2}, i.e., Platinum on Carbon cloth is loaded. To avoid leakages of methyl alcohol solutions, PTFE fabricated gaskets are used in between sub-assemblies joining of the DMFC system. An MEA of 5 cm by 5 cm active area is used in the cell and the cell has experimented with 5 molar methanol concentration. The necessary compact clamping is achieved using bolting of M8 size and the uniform torquing is ensured using a handheld torque wrench to a pre-set value of 5 N-m.

Experiment Methodology

Nickel current collector material has been selected for experimentation as it has good electrical conductivity and is better compatible with methanol solution. PTFE gaskets of 0.11 mm thickness are chosen for fabrication of controlling shapes and sizes of effective flow path area over MEA in the cell. The controlling shape of the circular opening in both equal areas and equal perimeters is chosen as a reference. The maximum size circular shape with a diameter of 50mm is cut in the PTFE gasket to accommodate in 50mm side square area of MEA. This circular shape has an effective area of 1963.5 mm^2 and a perimeter of 177.2 mm. Keeping this effective area and perimeter as controlling parameters, other shapes such as square, rectangle, and rhombus gaskets are used on either side of MEA at a time. Methyl alcohol solution with 5 molar concentration is used to perform this experiment. By changing current parameters (keeping fixed current values), voltage values are recorded from OCV to the lowest attainable reading. Performance characteristics of the cell are investigated using Polarization and power density curves at this5molar methyl alcohol concentration. The total experiment has been performed thrice to have consistency and repeatability in the noted readings. The arithmetic average reading of the noted values of the experiment conforming to generated voltage and corresponding current is referred for analyzing the cell performance characteristics.

Details of the gasket dynamic openings, such as circle having diameter 50 mm, rectangle length 50 mm X breadth 39.27 mm, square with 44.31 mm side and rhombus with 44.95 mm side in equal areas and the gasket openings such as circle having diameter 50 mm, rectangle length 50 mm X breadth 28.54 mm, square with 39.27 mm side and rhombus with 39.27 mm side in equal perimeters.

The setup of the gasket dynamic openings over the anode-side current collector and over MEA is shown in figure-3 and figure-4 respectively.

SUMMARY

Influence of effective Opening Shapes on cell performance

With equal areas

Polarization (Voltage versus Current density) curves and power density (Power density versus Current density) curves are drawn with the experiment values. It is observed that the cell using square shape has produced a maximum power density of 6.344 $mWcm^{-2}$ among the other shapes chosen that includes rhombus, rectangle, and circle. The circular shape is producing the least power density of 5.541 $mWcm^{-2}$. The better performance of the cell with square shape active area is because of equal circulation of reactants on the effective area. Figure-5 has shown the cell characteristics with various shapes of equal areas in gasket openings.

With equal perimeters

Polarization (Voltage versus Current density) curves and power density (Power density versus Current density) curves are drawn from the experimental results. It is observed that the cell with a rhombus shape has produced a maximum power density of 7.714 $mWcm^{-2}$ among the other shapes chosen that includes square, rectangle, and circle. The circular shape is producing the least power density of 5.541

mWcm^{-2}. The better performance of the cell with rhombus shape is due to less effective area among equal perimeters of selected shapes. Figure-6 has showed the cell characteristics with various shapes of equal perimeters in gasket openings.

Influence of effective Opening Shapes on Current Density

With equal areas

From the polarization (Voltage versus Current density) curves that are drawn using the experiment values, it is observed that the cell using square shape has produced a maximum current density of 65.2 mAcm^{-2} among the other shapes chosen that include rhombus, rectangle, and circle. The circular shape is producing the least current density of 55.0 mAcm^{-2}. The better performance of the cell with square shape active area is because of equal circulation of reactants on the effective area.

With equal perimeters

Polarization (Voltage versus Current density) curves are drawn from the experimental results, it is observed that the cell with rhombus shape has produced a maximum power density of 85.3 mAcm^{-2} among the other shapes chosen that include square, rectangle, and circle. The circular shape is producing the least power density of 55.0 mAcm^{-2}. The better performance of the cell with rhombus shape is due to less effective area among equal perimeters of selected shapes.

Comparison of gasket opening perimeter and area effective Opening Shapes

With equal areas

From the various shapes of equal areas, it is observed that the circular shape has the lowest opening perimeter and the rhombus has the largest perimeter. Followed by rhombus shape, rectangle and square shapes have fewer opening perimeters in the descending order.

With equal perimeters

From the various shapes of equal perimeters, it is observed that the rhombus shape has the lowest opening area and the circle has the largest area. Therefore, it is measured from the geometries of various shapes that rhombus, rectangle, square and circle have increasing gasket opening areas. Hence power density is influenced by the area of the opening in equal perimeter shapes.

Influence of gasket opening area to perimeter ratio (A/P) on effective gasket opening Shapes

With equal areas

From the various shapes of equal areas, it is observed that the ratio of gasket opening area to the perimeter is also influencing the performance. The circular shape has the highest opening area to the perimeter and the rhombus has the least value. From the performance graphs, it is revealed that the circular one has the least performance due to the influence of the highest gasket opening area to perimeter ratio.

With equal perimeters

From the various shapes of equal perimeters, it is observed that the ratio of gasket opening area to the perimeter is also influencing the performance. The circular shape has the highest opening area to the perimeter and the rhombus has the least value. From the performance graphs, it is revealed that the circular one has the least performance due to the influence of the highest gasket opening area to perimeter ratio and the rhombus shape has the highest performance due to the lowest gasket opening area to perimeter ratio. It is revealed that the gasket opening areas with various shapes such as rhombus, rectangle, square, and circle have better performance with decreasing gasket opening area to perimeter ratio on effective opening shapes.

Influence of Current collector's opening perimeter and opening area on gasket effective Opening Shapes

With equal areas

From the selected shapes in equal areas, it is observed that the current collector opening perimeter is influencing the performance of the cell. The rectangular shape has the highest opening perimeter and the square has the least value. From the performance graphs, it is revealed that the square shape has the highest performance due to the influence of the least opening perimeter on the current collector.

Similarly, from the various shapes of equal areas, it is observed that the current collector opening area is also influencing the performance. The square shape has the largest opening area and the rectangle has the least value. From the performance graphs, it is identified that the square shape has the highest performance due to the influence of the least opening area on the current collector.

With equal perimeters

From the identified shapes of equal perimeters, it is observed that the current collector opening perimeter is also influencing the cell performance. Rhombus shape has the least opening perimeter and circular shape has largest opening perimeter on the current collector. From the performance graphs, it is observed that the circular one has the least performance due to the largest opening perimeter on current collectors and the rhombus shape has the highest performance due to the lowest opening perimeter on current collectors.

Similarly, from the various shapes of equal perimeters, it is observed that the current collector opening area is also influencing the performance. Rhombus shape has the least opening area and the circle has the largest opening area on current collectors. From the performance graphs, it is revealed that the rhombus shape has the highest performance due to the influence of the least opening area on the current collector.

Influence of Current collector's opening area to opening perimeter ratio

With equal areas

From the various shapes of equal areas, it is observed that the square shape has the largest ratio of current collector opening area to the perimeter, whereas the rectangle has the least ratio. Besides other influencing factors, the square shape is showing better performance with the largest ratio of current collector opening area to the perimeter.

With equal perimeters

From the different shapes of equal perimeters, it is observed that the rhombus shape has the least current collector opening area to perimeter ratio and the rectangle has the largest ratio. Besides other consideration factors, rhombus shape is showing the better performance with the least ratio of current collector opening area to the perimeter.

Influence of Current collector's opening ratio

With equal areas

From the assorted shapes in equal areas, it is observed that the square shape has the largest current collector opening ratio and the rectangle has the least ratio. Besides other factors, square shape is showing better performance with the largest ratio of current collector opening ratio.

With equal perimeters

From the various shapes of equal perimeters, it is observed that the rhombus shape has the least current collector opening ratio and the rectangle has the largest ratio. Besides other factors, rhombus shape is showing the better performance with the least ratio of current collector opening.

Comparison of Power and Power Density on Opening Shape sizes

Power generated Vs various sizes and shapes

From the assorted sizes of selected shapes, a comparison study on size factor (in terms of area of the dynamic zone) with Power produced is carried out. It is observed that the higher the effective MEA area, the higher the power produced by the cell for a given shape.

Power Density produced Vs various sizes and shapes

From the various sizes of selected shapes, a comparison study on size factor (in terms of area of the dynamic zone) with Power density produced is carried out. It is observed that the higher the effective MEA area, the lower the power produced by the cell for a given shape. Hence the performance of the cell lowers with enhance in the size of the dynamic zone.

Figure-7 and figure-8 have shown the comparison of power and power density respectively produced against the gasket opening area with various shapes and sizes.

OUTCOME

Our Invention is related to an Experimental Investigation on Performance of Passive Direct Methanol Fuel Cell using dynamic active regions of Membrane Electrode Assembly with Nickel Current Collectors. Comparison of the shapes such as square, rectangular, rhombus, and circular active area forms and by using Nickel as current collector material is selected for evaluation. In the equal areas with gasket opening area of 1963.5 (+/-0.2) mm^2, among circular, rhombus, rectangular and square shapes chosen for investigation on the performance of cell at 5 molar methanol concentration, square shape opening consisting of the perimeter of 177.2 mm has developed highest power density of 6.344 $mWcm^{-2}$with a corresponding maximum current density of 65.2 $mAcm^{-2}$. In the equal perimeters of 157.1 (+/- 0.2) mm of gasket openings, among circular, rhombus, rectangular and square shapes chosen for investigation on the performance of cell at 5 molar methanol concentration, Rhombus shape opening consisting of area 1400 mm^2has developed the highest power density of 7.714 $mWcm^{-2}$ with a corresponding maximum current density of 85.3 $mAcm^{-2}$.

1. According to claim 1# the invention is related to an Experimental Investigation on Performance of Passive Direct Methanol Fuel Cell using dynamic active regions of Membrane Electrode Assembly with Nickel Current Collectors In the large active areas, methanol consumption is slow because the current density is lower when compared to the smaller active areas. As a result, the surface-specific concentration on the ion-conducting membrane is being more, making a higher crossover of methanol over bigger active zones.
2. According to claim 1,2# the invention is the experimental investigation demonstrated that the size of the dynamic zone significantly affects the performance of the cell and its performance can be improved with the expansion in dynamic region size. In the vast dynamic regions, the current density is lower compared to the smaller dynamic zones where the methanol is expended gradually. So, the local concentration on the cell membrane is more and brings further fuel cross-over.
3. According to claim 1,2,3# the invention is the exothermic response that is involved on penetrated fuel and oxidant on cathode impetus generates more heat, which prompts an elevated working temperature. This elevated working temperature brings about superior electrochemical kinetics of fuel oxidation and a decrease in oxygen response, leading to superior cell performance.
4. According to claim 1,2,3,4# the invention is the internal obstruction decreases with expanding the size of the dynamic zone. This reality is good with the accomplished outcome. Even though the cell performance can be enhanced by expanding the size of the dynamic zone, with the research outcome as on date, it is revealed that tiny dynamic zones have its focal points, for example, more slow charge transfer obstruction, lesser flooding, even distribution of methanol dispersion, increasing of effective current gathering, minor expenditure on fabrication, and further, it is small in size and weight.
5. According to claim 1,2,3,5# the invention is from the experiment that cell performance depends on the combined effect of (i) gasket opening shape, opening area, and its perimeter; (ii) the free open ratio and opening perimeter on the current collector

plates. Comparison study referring to power and power density on dynamic section shows the influence of shapes and sizes of the geometry of effective MEA regions.

6. According to claim 1,2,3,6# the invention is that the fuel cell performance is a function of membrane-active area size and shapes; current collectors opening area size and shapes. The experimental results indicate that with equal areas, the cell performance is best using square shape compared to circle, rhombus, and rectangle. However, in the equal perimeters, the cell performance is best using square shape compared to circle, rhombus, and rectangle.

REFERENCES

[1] Abhigna, B.S., Chakraborty, S., Kumar, V., Medha: A collaborative E-concept framework, Proceedings of the 2018 International Conference On Communication, Computing and Internet of Things, IC3IoT 2018, 2019, pp. 117–122, 8668127

[2] Abidin, S., Vadi, V.K., Rana, A., On Confidentiality, Integrity, Authenticity, and Freshness (CIAF) in WSN, Advances in Computer, Communication and Computational Sciences, 2020, pp 87-97

[3] Ahay, A., Choudhury, T., Kumar, V., Hierarchical Evolution of Digital Games, Proceedings of 2019 International Conference on Computational Intelligence and Knowledge Economy, ICCIKE 2019, 2019, pp. 381–386, 9004378

[4] Area: Communication, Frequency modulation, Networks, wireless communication

[5] Area: Communication, Frequency modulation, Networks, wireless communication

[6] Area: Communication, Frequency modulation, Networks, wireless communication

[7] Area: Communication, Frequency modulation, Networks, wireless communication

[8] Arora, M., Kumar, V., Dave, M., Hybrid cloud infrastructure for load balancing and scaling with vm path optimization using ant colony optimization, Journal of Advanced Research in Dynamical and Control Systems, 2020, 12(5 Special Issue), pp. 660–674

[9] Arora, M., Kumar, V., Dave, M., Task scheduling in cloud infrastructure using optimization technique genetic algorithm, Proceedings of the World Conference on Smart Trends in Systems, Security and Sustainability, WS4 2020, 2020, pp. 788–793, 9210303

[10] Bansal, R., Gupta, A., Singh, R., Nassa, V.K., Role and impact of digital technologies in E-learning amidst COVID-19 pandemic, Proceedings - 2021 4th International Conference on Computational Intelligence and Communication Technologies, CCICT 2021, 2021, pp. 194–202

[11] Bansal, R., Obaid, A.J., Gupta, A., Singh, R., Pramanik, S., Impact of Big Data on Digital Transformation in 5G Era, Journal of Physics: Conference Seriesthis link is disabled, 2021, 1963(1), 012170

[12] Chauhan, S.S., Khare, A.K., Kinematic analysis of the ABB IRB 1520 industrial robot using roboanalyzer software, Evergreenthis link is disabled, 2020, 7(4), pp. 510–518

[13] Chauhan, S.S., Khare, A.K., Method of Obtaining Planar State of Stress Using Mohr's Circle—Some Typical Cases, Lecture Notes in Mechanical Engineering, 2022, pp. 33–38

[14] Goel, S., Tripathi, A., Choudhury, T., Kumar, V. Parkinson's Disease Detection using Soft Computing Technique, Proceedings of the 2019 8th International Conference on System Modeling and Advancement in Research Trends, SMART 2019, 2020, pp. 3–6, 9117384

[15] Govindarajan, V., Sivakumar, R., Patil, P.P., Kannan, M., Ramesh, B., Effect of Tungsten Carbide Addition on the Microstructure and Mechanical Behavior of Titanium Matrix Developed by Powder Metallurgy Route, Advances in Materials Science and Engineeringthis link is disabled, 2022, 2022, 2266951

[16] Gupta, A., Singh, R., Nassa, V.K., Sharma, P., Koti, K., Investigating Application and Challenges of Big Data Analytics with Clustering, 2021 International Conference on Advancements in Electrical, Electronics, Communication, Computing and Automation, ICAECA 2021, 2021

[17] Jain, P., Vijay, S., & Gupta, S. C. (2009). ATM Network QoS Parameters and Performance Analysis using OPNET Simulator. International Journal of Recent Trends in Engineering, 2(4), 228.

[18] Jain, P., Vijay, S., & Gupta, S. C. Performance Analysis & QoS Guarantee in ATM Networks. Global Journal of computer Science and Technology, 131-136.

[19] Jain, S., Choudhury, T., Kumar, V., Kumar, P., Detecting malware and analysing using sandbox evasion, Proceedings of the 2018 International Conference On Communication, Computing and Internet of Things, IC3IoT 2018, 2019, pp. 111–116, 8668149

[20] Kala, S., Nandan, H., Sharma, P., Shadow and weak gravitational lensing of a rotating regular black hole in a non-minimally coupled Einstein-Yang-Mills theory in the presence of plasma, European Physical Journal Plusthis link is disabled, 2022, 137(4), 457

[21] Kandpal, H., Kumar, A., Malik, A., Morphometric analysis and prioritization of sub-watersheds in Bino watershed, Uttarakhand: A remote sensing and GIS perspective, Indian Journal of Ecology, 2018, 45(3), pp. 475–478

[22] Kannan, M., Singh, S., Prasad, R.R., Synthetic Methods to Obtain Calcia-Stabilized Zirconia Powders—A Review, Lecture Notes in Mechanical Engineeringthis link is disabled, 2021, pp. 405–416

[23] Kukreti, A., Painolli, A.K., Rana, N., Where do we stand: factors affecting sustainable development, Procedia Environmental Science, Engineering and Management, Volume 8, Issue 2, 2021, Pages 583-589

[24] Kulkarni, S & Chima, P 2021 'Challenges faced by UK university students due to thcoronavirus crisis in the Higher Education' Preprints.org. https://doi.org/10.20944/preprints202102.0192.v1

[25] Kulkarni, S & Sodre, JR 2017, An Assessment of Lightweight Vehicles to Reduce Greenhouse Gas Emissions with Focus on Magnesium. in Thoughts and Refections on the use of Magnesium. Birmingham City University, pp. 84-91.

[26] Kulkarni, S, Chapman, C & Shah, H 2016 'Computational Fluid Dynamics (CFD) Mesh Independency Study of A Straight Blade Horizontal Axis Tidal Turbine' Preprints.org. https://doi.org/10.20944/preprints201608.0008.v1

[27] Kulkarni, S, Chapman, C, Shah, H & Edwards, D 2018, 'A computational design method for bio-mimicked horizontal axis tidal turbines', International Journal of Building Pathology and Adaptation, vol. 36, no. 2, pp. 188-209. https://doi.org/10.1108/IJBPA-06-2017-0029

[28] Kulkarni, S, Chapman, C, Shah, H, Parn, E & Edwards, DJ 2018, 'Designing an efficient tidal turbine blade through bio-mimicry: A systematic review', Journal of Engineering, Design and Technology, vol. 16, no. 1, pp. 101-124. https://doi.org/10.1108/JEDT-08-2017-0077

[29] Kulkarni, S, Chapman, C, Shah, H, Parn, E & Edwards, DJ 2018, 'Design Study of Horizontal Axis Tidal Turbine Blade', Mindanao Journal of Science and Technology. https://www.open-access.bcu.ac.uk/10089/

[30] Kulkarni, S, Edwards, DJ, Chapman, C, Hosseini, MR & Owusu-Manu, D-G 2019, 'A preliminary mechanical design evaluation of the Wikispeed car: for light-weighting implications', Journal of Engineering, Design and Technology, vol. 17, no. 1, pp. 230-249. https://doi.org/10.1108/JEDT-09-2018-0154

[31] Kulkarni, S, Edwards, DJ, Parn, E, Chapman, C, Aigbavboa, CO & Cornish, R 2018, 'Evaluation Of Vehicle Lightweighting To Reduce Greenhouse Gas Emissions With Focus On Magnesium Substitution', Journal of Engineering, Design and Technology, vol. 16, no. 6, pp. 869-888. https://doi.org/10.1108/JEDT-03-2018-0042

[32] Kulkarni, SS, Wang, L, Golsby, N & Lander, M 2021, 'Fluid-structure interaction based optimisation in tidal turbines: A perspective review', Journal of Ocean Engineering and Science. https://doi.org/10.1016/j.joes.2021.09.017

[33] Malik, A., Kumar, A., Kandpal, H., Morphometric analysis and prioritization of sub-watersheds in a hilly watershed using weighted sum approach, Arabian Journal of Geosciencesthis link is disabled, 2019, 12(4), 118

[34] Mohanakurup, V.K., Baghela, V.S., Kumar, S., Srivastava, P.K., Doohan, N.V., Soni, M., Awa, H., 5G Cognitive Radio Networks Using Reliable Hybrid Deep Learning Based on Spectrum Sensing, Green Internet of Things in 5G and Beyond, Hindawi, Volume 2022

[35] Mohta, M., Soni, N., Choudhury, T., Kumar, V., Market Analysis on Li-Fi Technology, Proceedings of the 2019 8th International Conference on System Modeling and Advancement in Research Trends, SMART 2019, 2020, pp. 139–143, 9117427

[36] Nagwan, H., Rana, N., Kumar, A., Tewari, P.C., Performance and emission characteristics of diesel engine using Karanja oil methyl ester and Eucalyptus oil fuel blends, Procedia Environmental Science, Engineering and Management, Volume 6, Issue 4, 2019, Pages 553-563

[37] Painoli, A.K., Bansal, R., Singh, R., Kukreti, A., Impact of digital marketing on the buying behavior of youth with special reference to Uttarakhand state, Big Data Analytics for Improved Accuracy, Efficiency, and Decision Making in Digital Marketing, 2021, pp. 162–182

[38] Parag, J. (2011). Hybrid mixed handover for call blocking probability analysis in wireless ATM network. Journal of Engineering and Computer Innovations, 2(4), 68-77.

[39] Sharma, C., Singh, M.P., Chaudhary, S., Cost Analysis of Solar-Wind Based Hybrid Renewable Energy System, Lecture Notes in Electrical Engineeringthis link is disabled, 2021, 700, pp. 2733–2746

[40] Sharma, S., Chandra, U., & Jain, P. (2017). A literature survey on automation of test data generation for branch coverage testing using genetic algorithm. International Journal of Computational Intelligence Research, 13(6), 1521-1531.

[41] Varshney, R., Jain, P., & Vijay, S. (2018, September). Massive MIMO Systems In Wireless Communication. In 2018 2nd International Conference on Micro-Electronics and Telecommunication Engineering (ICMETE) (pp. 39-44). IEEE.

Chapter - 08

Increase Speed Using Machine Learning

Dr. Anil W. Kale

Mahatma Gandhi Mission's College of Engineering and Technology, India.
Email: anil5474@gmail.com

Dr. Nandkishor P Karlekar

MGMCET Kamothe Navimumbai, India. Email: npkarlekar.mgmcet@gmail.com

Dr. Vaibhav Narawade

Ramrao Adik Institute of Technology, India. Email:vnarawade@gmail.com

Dr. Sanjay B. Waykar

Mahatma Gandhi Mission's College of Engineering and Technology, India.
Email:sbwaykar@gmail.com

ABSTRACT

Our Invention "increase speed using Machine Learning." is a Outflow free compacted air fueled energy framework can be utilized as the principle power source or as a helper power unit in vehicular transportation with benefits of zero fossil fuel byproducts and further developed the general energy proficiency of the coordinated energy framework. This work introduced an itemized mechanical improvement of compacted air energy frameworks. The investigations on compacted air controlled powertrain in transport area are summed up including the plan of new valve advances, model turns of events, and joining of the framework. Besides, packed air based crossover advances utilizing different pneumatic hybridization strategies are thoroughly introduced meaning to give inside and out understanding on the benefits and limits of various pneumatic hybridization. The open doors and difficulties for the packed air based innovation in transport application are talked about. It tends normal the change of energy frameworks to a cleaner and more economical future would advance the mechanical turn of events and execution of Zero-Emission packed air arrangements.

SUMMARY

As one of the potential advances conceivably accomplishing zero emanations target, packed air controlled impetus frameworks for transport application have drawn in expanding research centers. Then again, the packed air energy unit can be coordinated with regular Internal Combustion Engine (ICE) shaping a mixture framework.

The crossover powertrain framework can be perceived as a progress vehicle drive framework between the fossil-fuelled burning motor and zero-outflow impetus framework, for example, electric power unit or packed air controlled unit. The crossover powertrain is right now quite possibly the most achievable answer for further develop mileage and decrease the emanations with somewhat in general low capital expense contrasted with the electric powertrains utilizing Lithium-particle batteries.

A crossover powertrain comprises of two power units, including a gas powered motor and an assistant power unit with a perfect energy source like a battery or packed air unit. The cross breed powertrain empowers customary fossil-fuelled ICE to be worked at upgraded conditions with lower fuel utilization and diminished contamination outflows.

OBJECTIVE

- The objective of the invention is to reduce drag force acting on vehicle using blower fan.
- The other objective of the invention is to increase speed and reduce fuel consumption (increase mileage) of vehicle
- The other objective of the invention is to of green energy (wind force) can reduce fuel consumption and avoid air pollution.
- The other objective of the invention is to "Airblower fan over car (Vehicle) to increase speed and mileage of vehicle" is an Outflow free compacted air fueled energy framework can be utilized as the principle power source or as a helper power unit in vehicular transportation with benefits of zero fossil fuel byproducts and further developed the general energy proficiency of the coordinated energy framework.

- The other objective of the invention is to work introduced an itemized mechanical improvement of compacted air energy frameworks. The investigations on compacted air controlled powertrain in transport area are summed up including the plan of new valve advances, model turns of events, and joining of the framework.
- The other objective of the invention is to packed air based crossover advances utilizing different pneumatic hybridization strategies are thoroughly introduced meaning to give inside and out understanding on the benefits and limits of various pneumatic hybridization. The open doors and difficulties for the packed air based innovation in transport application are talked about.
- The other objective of the invention is too normal the change of energy frameworks to a cleaner and more economical future would advance the mechanical turn of events and execution of Zero-Emission packed air arrangements.

SUMMARY

Before the creation of the fossil-fuelled Internal Combustion Engine (ICE), packed air powertrain was generally utilized as the power source in train transportation for over 50 years because of its straightforwardness, security and minimal expense. Train motors driven by packed air were broadly evolved during the 1880s and 1890s by organizations like General Herman Haupt. In 1886, Louis Mekarski fostered the Mekarski train framework.

The utilizing a solitary chamber packed air motor, which a boiling water tank warmer was utilized to warm compacted air before the extension in the motor chamber [6]. Robert Hardie further fostered the idea by embracing a high level re-warming framework on a one-stage extension motor with the capacity of recovering motor brake energy.

A two-stage air-controlled motor called Hoadley-Knight framework was produced to broaden the working distance of the train prior to re-energizing the framework. The Hardie framework and Hoadley-Knight framework were both worked in New York USA during 1880-1900, and the packed air was provided from a 1500 torque steam-fueled air blower station.

During that time, the air-driven train was additionally utilized in coal mineshafts. The packed air-driven train won't create tremendous hotness or flash during the activity, which wipe out the gamble of blast and subsequently more secure than its primary rival the Internal Combustion Engine.

BRIEF DESCRIPTION OF THE DIAGRAM

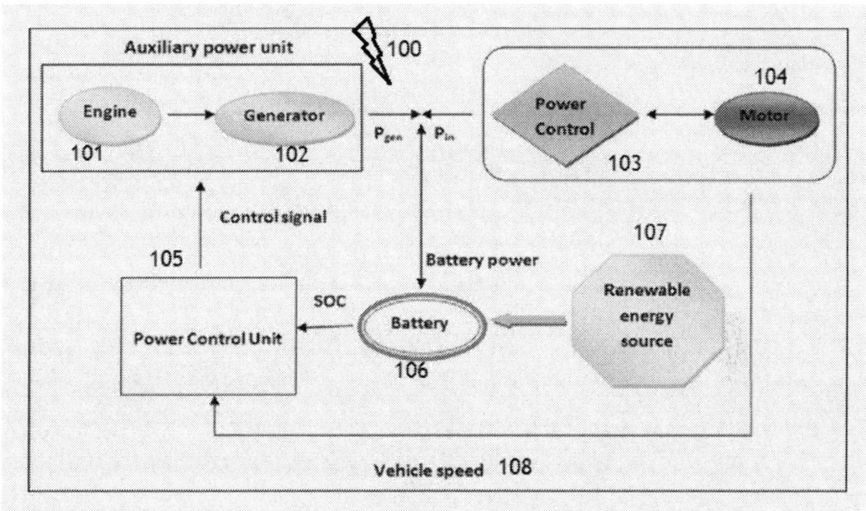

Fig. 1: Air Blower fan for car (Vehicle) to Increase Speed and Mileage of Vehicle Flow Chart.

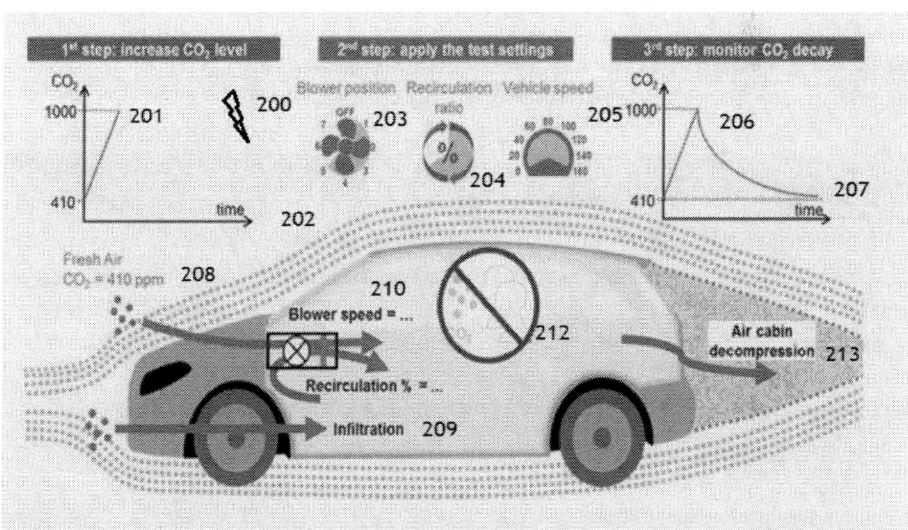

Fig. 2: Airblower fan over car (Vehicle) to increase speed and mileage of vehicle, Flow chart.

DESCRIPTION

Working principles of compressed air powered engines

In a packed air drive framework, the energy is typically changed over by the motor into mechanical power through the development of compacted air in the chamber. Shows the functioning system of a typical packed air controlled motor. The responding cylinder structure is like that of a traditional gas powered motor with the exception of that packed air controlled motor has no fuel injector or flash attachment.

The functioning pattern of the motor is made out of 2 strokes, known as development and exhaust. It should be noticed that the admission cycle occurs toward the start of the development stroke. The admission valve typically opens when the cylinder moves to the Top Dead Center (TDC), and the compacted wind currents into the chamber from the air tank.

After the admission valve is shut, the packed air caught in the chamber keeps on extending until the cylinder arrives at the Bottom Dead Center (BDC). The exhaust valve opens after the development interaction, and the packed air is siphoned out of the chamber with the upwards development.

Design and optimisation of the valve system

The energy change cycles of the packed air decide the presentation of the compacted air fuelled motor. It is hence basic to control and advance the admission and exhaust cycle to decrease the stream misfortunes. Many investigations were directed on the plan, re-enactment and trial of the valve framework to accomplish an improved presentation and proficiency of the compacted air fuelled motor.

For instance: the tension decrease cycle of packed air during the admission interaction and reasoned that the energy proficiency could be improved by 15-40% while taking on pressure decrease by expander as opposed to choking depressurisation, and the energy transformation capacity of compacted air could be expanded by 10-60% assuming a two-stage pressure decrease was embraced.

Furthermore, high tension on/off valve with its overall control calculation was intended for the packed air controlled vehicle framework. Proposed a rotational valve framework driven by a sprocket wheel associated with the driving rod. A numerical model of the admission and fumes process was laid out in light of the rotational valve framework, and the impact of the valve timing on the exhibition of compacted air fuelled motor was examined. The outcomes showed that the most advanced valve timing is that the admission valve opened at the Top Dead Center (TDC).

Planned a valve framework with a polynomial cam profile and re-enacted the impacts of the half wrap point and the most extreme valve lift to the completion coefficient. It was presumed that planned cam profile should utilize the base valve lift to get the most extreme completion coefficient.

Applied the symmetrical plan with dark connection examination to improve the valve timing boundaries of the packed air controlled motor and proposed the ideal upsides of cam rise point and cam bring point back. Fostered a numerical model of a compacted air controlled motor and approved the reproduction in a trial model. A strain remunerated admission valve was planned and acquainted with concentrate on the primary boundaries plan technique.

OUTCOME

Our Invention "Airblower fan over car (Vehicle) to increase speed and mileage of vehicle" is an Outflow free compacted air fueled energy framework can be utilized as the principle power source or as a helper power unit in vehicular transportation with benefits of zero fossil fuel byproducts and further developed the general energy proficiency of the coordinated energy framework. This work introduced an itemized mechanical improvement of compacted air energy frameworks. The investigations on compacted air controlled powertrain in transport area are summed up including the plan of new valve advances, model turns of events, and joining of the framework. Besides, packed air based crossover advances utilizing different pneumatic hybridization strategies are thoroughly introduced meaning to give inside and out understanding on the benefits and limits of various pneumatic hybridization. The open doors and difficulties for the packed air based innovation in transport application are talked about. It tends normal the change of energy frameworks to a cleaner and more economical future would advance the mechanical turn of events and execution of Zero-Emission packed air arrangements.

1. According to claim1# the invention is to a "Airblower fan over car (Vehicle) to increase speed and mileage of vehicle" is an Outflow free compacted air fueled energy framework can be utilized as the principle power source or as a helper power unit in vehicular transportation with benefits of zero fossil fuel

byproducts and further developed the general energy proficiency of the coordinated energy framework.

2. According to claim1, 2, 3# the invention is to a work introduced an itemized mechanical improvement of compacted air energy frameworks. The investigations on compacted air controlled powertrain in transport area are summed up including the plan of new valve advances, model turns of events, and joining of the framework.
3. According to claim1,2, 3# the invention is to a packed air based crossover advances utilizing different pneumatic hybridization strategies are thoroughly introduced meaning to give inside and out understanding on the benefits and limits of various pneumatic hybridization. The open doors and difficulties for the packed air based innovation in transport application are talked about.
4. According to claim1,2,3,4# the invention is to a normal the change of energy frameworks to a cleaner and more economical future would advance the mechanical turn of events and execution of Zero-Emission packed air arrangements.

REFERENCES

[1] Abhigna, B.S., Chakraborty, S., Kumar, V., Medha: A collaborative E-concept framework, Proceedings of the 2018 International Conference On Communication, Computing and Internet of Things, IC3IoT 2018, 2019, pp. 117–122, 8668127

[2] Abidin, S., Vadi, V.K., Rana, A., On Confidentiality, Integrity, Authenticity, and Freshness (CIAF) in WSN, Advances in Computer, Communication and Computational Sciences, 2020, pp 87-97

[3] ahay, A., Choudhury, T., Kumar, V., Hierarchical Evolution of Digital Games, Proceedings of 2019 International Conference on Computational Intelligence and Knowledge Economy, ICCIKE 2019, 2019, pp. 381–386, 9004378

[4] Area: Communication, Frequency modulation, Networks, wireless communication

[5] Area: Communication, Frequency modulation, Networks, wireless communication

[6] Area: Communication, Frequency modulation, Networks, wireless communication

[7] Area: Communication, Frequency modulation, Networks, wireless communication

[8] Arora, M., Kumar, V., Dave, M., Hybrid cloud infrastructure for load balancing and scaling with vm path optimization using ant colony optimization,

Journal of Advanced Research in Dynamical and Control Systems, 2020, 12(5 Special Issue), pp. 660–674

[9] Arora, M., Kumar, V., Dave, M., Task scheduling in cloud infrastructure using optimization technique genetic algorithm, Proceedings of the World Conference on Smart Trends in Systems, Security and Sustainability, WS4 2020, 2020, pp. 788–793, 9210303

[10] Bansal, R., Gupta, A., Singh, R., Nassa, V.K., Role and impact of digital technologies in E-learning amidst COVID-19 pandemic, Proceedings - 2021 4th International Conference on Computational Intelligence and Communication Technologies, CCICT 2021, 2021, pp. 194–202

[11] Bansal, R., Obaid, A.J., Gupta, A., Singh, R., Pramanik, S., Impact of Big Data on Digital Transformation in 5G Era, Journal of Physics: Conference Seriesthis link is disabled, 2021, 1963(1), 012170

[12] Chauhan, S.S., Khare, A.K., Kinematic analysis of the ABB IRB 1520 industrial robot using roboanalyzer software, Evergreenthis link is disabled, 2020, 7(4), pp. 510–518

[13] Chauhan, S.S., Khare, A.K., Method of Obtaining Planar State of Stress Using Mohr's Circle—Some Typical Cases, Lecture Notes in Mechanical Engineering, 2022, pp. 33–38

[14] Goel, S., Tripathi, A., Choudhury, T., Kumar, V. Parkinson's Disease Detection using Soft Computing Technique, Proceedings of the 2019 8th International Conference on System Modeling and Advancement in Research Trends, SMART 2019, 2020, pp. 3–6, 9117384

[15] Govindarajan, V., Sivakumar, R., Patil, P.P., Kannan, M., Ramesh, B., Effect of Tungsten Carbide Addition on the Microstructure and Mechanical Behavior of Titanium Matrix Developed by Powder Metallurgy Route, Advances in Materials Science and Engineeringthis link is disabled, 2022, 2022, 2266951

[16] Gupta, A., Singh, R., Nassa, V.K., Sharma, P., Koti, K., Investigating Application and Challenges of Big Data Analytics with Clustering, 2021 International Conference on Advancements in Electrical, Electronics, Communication, Computing and Automation, ICAECA 2021, 2021

[17] Jain, P., Vijay, S., & Gupta, S. C. (2009). ATM Network QoS Parameters and Performance Analysis using OPNET Simulator. International Journal of Recent Trends in Engineering, 2(4), 228.

[18] Jain, P., Vijay, S., & Gupta, S. C. Performance Analysis & QoS Guarantee in ATM Networks. Global Journal of computer Science and Technology, 131-136.

[19] Jain, S., Choudhury, T., Kumar, V., Kumar, P., Detecting malware and analysing using sandbox evasion, Proceedings of the 2018 International Conference On Communication, Computing and Internet of Things, IC3IoT 2018, 2019, pp. 111–116, 8668149

[20] Kala, S., Nandan, H., Sharma, P., Shadow and weak gravitational lensing of a rotating regular black hole in a non-minimally coupled Einstein-Yang-Mills theory in the presence of plasma, European Physical Journal Plusthis link is disabled, 2022, 137(4), 457

[21] Kandpal, H., Kumar, A., Malik, A., Morphometric analysis and prioritization of sub-watersheds in Bino watershed, Uttarakhand: A remote sensing and GIS perspective, Indian Journal of Ecology, 2018, 45(3), pp. 475–478

[22] Kannan, M., Singh, S., Prasad, R.R., Synthetic Methods to Obtain Calcia-Stabilized Zirconia Powders—A Review, Lecture Notes in Mechanical Engineeringthis link is disabled, 2021, pp. 405–416

[23] Kukreti, A., Painolli, A.K., Rana, N., Where do we stand: factors affecting sustainable development, Procedia Environmental Science, Engineering and Management, Volume 8, Issue 2, 2021, Pages 583-589

[24] Kulkarni, S & Chima, P 2021 'Challenges faced by UK university students due to thcoronavirus crisis in the Higher Education' Preprints.org. https://doi.org/10.20944/preprints202102.0192.v1

[25] Kulkarni, S & Sodre, JR 2017, An Assessment of Lightweight Vehicles to Reduce Greenhouse Gas Emissions with Focus on Magnesium. in Thoughts and Refections on the use of Magnesium. Birmingham City University, pp. 84-91.

[26] Kulkarni, S, Chapman, C & Shah, H 2016 'Computational Fluid Dynamics (CFD) Mesh Independency Study of A Straight Blade Horizontal Axis Tidal Turbine' Preprints.org. https://doi.org/10.20944/preprints201608.0008.v1

[27] Kulkarni, S, Chapman, C, Shah, H & Edwards, D 2018, 'A computational design method for bio-mimicked horizontal axis tidal turbines', International Journal of Building Pathology and Adaptation, vol. 36, no. 2, pp. 188-209. https://doi.org/10.1108/IJBPA-06-2017-0029

[28] Kulkarni, S, Chapman, C, Shah, H, Parn, E & Edwards, DJ 2018, 'Designing an efficient tidal turbine blade through bio-mimicry: A systematic review', Journal of Engineering, Design and Technology, vol. 16, no. 1, pp. 101-124. https://doi.org/10.1108/JEDT-08-2017-0077

[29] Kulkarni, S, Chapman, C, Shah, H, Parn, E & Edwards, DJ 2018, 'Design Study of Horizontal Axis Tidal Turbine Blade', Mindanao Journal of Science and Technology. https://www.open-access.bcu.ac.uk/10089/

[30] Kulkarni, S, Edwards, DJ, Chapman, C, Hosseini, MR & Owusu-Manu, D-G 2019, 'A preliminary mechanical design evaluation of the Wikispeed car: for light-weighting implications', Journal of Engineering, Design and Technology, vol. 17, no. 1, pp. 230-249. https://doi.org/10.1108/JEDT-09-2018-0154

[31] Kulkarni, S, Edwards, DJ, Parn, E, Chapman, C, Aigbavboa, CO & Cornish, R 2018, 'Evaluation Of Vehicle Lightweighting To Reduce Greenhouse Gas Emissions With Focus On Magnesium Substitution', Journal of Engineering, Design and Technology, vol. 16, no. 6, pp. 869-888. https://doi.org/10.1108/JEDT-03-2018-0042

[32] Kulkarni, SS, Wang, L, Golsby, N & Lander, M 2021, 'Fluid-structure interaction based optimisation in tidal turbines: A perspective review', Journal of Ocean Engineering and Science. https://doi.org/10.1016/j.joes.2021.09.017

[33] Malik, A., Kumar, A., Kandpal, H., Morphometric analysis and prioritization of sub-watersheds in a hilly watershed using weighted sum approach, Arabian Journal of Geosciencesthis link is disabled, 2019, 12(4), 118

[34] Mohanakurup, V.K., Baghela, V.S., Kumar, S., Srivastava, P.K., Doohan, N.V., Soni, M., Awa, H., 5G Cognitive Radio Networks Using Reliable Hybrid Deep Learning Based on Spectrum Sensing, Green Internet of Things in 5G and Beyond, Hindawi, Volume 2022

[35] Mohta, M., Soni, N., Choudhury, T., Kumar, V., Market Analysis on Li-Fi Technology, Proceedings of the 2019 8th International Conference on System Modeling and Advancement in Research Trends, SMART 2019, 2020, pp. 139–143, 9117427

[36] Nagwan, H., Rana, N., Kumar, A., Tewari, P.C., Performance and emission characteristics of diesel engine using Karanja oil methyl ester and Eucalyptus oil fuel blends, Procedia Environmental Science, Engineering and Management, Volume 6, Issue 4, 2019, Pages 553-563

[37] Painoli, A.K., Bansal, R., Singh, R., Kukreti, A., Impact of digital marketing on the buying behavior of youth with special reference to Uttarakhand state, Big Data Analytics for Improved Accuracy, Efficiency, and Decision Making in Digital Marketing, 2021, pp. 162–182

[38] Parag, J. (2011). Hybrid mixed handover for call blocking probability analysis in wireless ATM network. Journal of Engineering and Computer Innovations, 2(4), 68-77.

[39] Sharma, C., Singh, M.P., Chaudhary, S., Cost Analysis of Solar-Wind Based Hybrid Renewable Energy System, Lecture Notes in Electrical Engineeringthis link is disabled, 2021, 700, pp. 2733–2746

[40] Sharma, S., Chandra, U., & Jain, P. (2017). A literature survey on automation of test data generation for branch coverage testing using genetic algorithm. International Journal of Computational Intelligence Research, 13(6), 1521-1531.

[41] Varshney, R., Jain, P., & Vijay, S. (2018, September). Massive MIMO Systems In Wireless Communication. In 2018 2nd International Conference on Micro-Electronics and Telecommunication Engineering (ICMETE) (pp. 39-44). IEEE.

CHAPTER - 09

Concrete Strength Monitoring using Machine Learning

Dr. Nandkishor P Karlekar

MGMCET Kamothe Navimumbai, India. Email: npkarlekar.mgmcet@gmail.com

Dr. Vaibhav Narawade

Ramrao Adik Institute of Technology, India. Email:vnarawade@gmail.com

Dr. Sanjay B. Waykar

Mahatma Gandhi Mission's College of Engineering and Technology, India. Email:sbwaykar@gmail.com

Dr. Anil W. Kale

Mahatma Gandhi Mission's College of Engineering and Technology, India. Email: anil5474@gmail.com

ABSTRACT

Our Invention "In-situ IoT Cloud Based Concrete Maturity-strength Measurement Systems for Real Time Concrete strength Monitoring and Prediction" is to a Wireless Embedded Maturity-Strength Measurement System (WE-MSMS) is designed and developed using digital DS18B20 Temperature Sensor array probe, microcontroller and Zig-Bee network with application software for maturity-strength measurement by non-destructive approach to estimate the in-situ concrete compressive strength of structure. A portable Wireless Temperature Sensor node with Zig-Bee network can be placed at construction site in the fresh concrete initially. Four-Sensor Nodes GUI Java platform based Application software is designed and used in the main server. In each sensor node there are eight- digital DS18B20 Temperature Sensors. For the measurement DS18B20 Temperature Sensor array probe embed into the fresh concrete, each DS18B20 Temperature Sensor with unique code is recorded and maintain in the database at main server. Thereafter every 5-minute or according to time setting till 28-days, time-temperature updated information will recorded in the central database using Zig-Bee network and IoT cloud. The designed system is useful for civil engineer to real-time prediction of in-situ concrete compressive strength at anytime and anywhere with the help of wireless technologies and IoT cloud which decide in-place grading of concrete in concrete construction work. Wireless maturity-strength measuring systems (WE-MSMS) can be implemented on different construction sites and connected with IoT cloud for real-time prediction of in-situ concrete compressive strength.

BACKGROUND

In traditional cube test method for compressive strength measurement of concrete mix design (IS 10262:2009). This is time consuming, wastage of concrete material and laborious job. The invented system provides the following solutions:

- In-situ IoT cloud based concrete maturity-strength measurement systems for real time concrete strength monitoring and prediction.
- It can be monitored concrete compressive strength by non-destructive approach and concrete maturity method.
- It can be maintained central database of concrete maturity-strength record in the main server of construction sites from all placed Sensor Nodes for further analysis.
- It can generate real time concrete maturity measurement based on IoT cloud as per the civil engineer's requirement within a second.
- It can also predict the real time concrete compressive strength to prevent failure of the concrete structure.
- It can also operate in dual power modes *either solar PV mode* or *electricity mode.*

PRIOR ART SEARCH

- Concrete is essential part of civil engineering for the construction of any architectural structure viz. building, parking structure, pavements, bridges, roads, runways, water storage dams, cement pipes and other purposes. During most of construction works it is found that there is a lack of understanding of exact concrete maturity-strength because one cannot measure maturity-strength in terms of any figure or by using any type of measuring instrument.

- This was the inspiration for undertaking the project work to develop embedded system that can exactly measure concrete maturity-strength in-place on experimental basis which can directly help civil engineering department to decide the grading of concrete that is compressive strength of concrete used in the construction site.
- There is a great importance of maturity-strength measurement for any type of concrete construction work. The most common and probable consequences of failure to maintain sufficient maturity-strength of any civil structure can cause dangerous results such as collapsing of entire structure or early visible cracks and leakages or bends in the concrete structure are seen due to premature construction work.
- The concrete in-place maturity-strength monitoring information is important to determine structural strength for serving at early age. In current scenario, the wireless temperature sensor node based embedded technology helps for monitoring early ages of concrete in-place maturity-strength. The concrete maturity method uses the time-temperature factor history (TTF) of concrete to estimate compressive strength development during curing time period.
- The designed and developed Embedded Wireless Maturity-Compressive Strength Measurement System (EW-MSMS) which is used for maturity-strength measurement consists of non-destructive method for estimating concrete compressive strength under investigation in the field during construction practice.
- This EW-MSMS plays an important role to collect, process, store, transmit and display the data/information related to concrete compressive strength (CCS). The specialty of EW-MSMS is that the collected information can be accessed at any time and also in-place with the help of wireless communication network and IoT cloud.

Advantages Over the Existing Art

1. Central system and real-time in-situ concrete-maturity measurement database control by non-destructive approach.
2. Wireless Temperature Sensor Nodes, IoT cloud based, portable and handy concrete-maturity measurement system can directly help civil engineers in remote area.
3. Time-temperature updated information recorded in the central system using Wireless Embedded Maturity-Strength Measurement System (WE-MSMS) by non-destructive approach and maturity method.
4. Dual Mode of Power: Solar PV and electricity.
5. Once it is activated, then it's on for 28-days for prediction of concrete compressive strength in real time.
6. Time-temperature updated information will be recorded in Real time and more accurately.
7. Maturity-Strength Reports of concrete construction using Nurse-Saul, Freiesleben Hansen and Pedersen and CEB-FIP mode are available within a click as per the civil engineers requirement.

OBJECTIVES

- The objective of the invention is to provide a "Real Time Concrete strength Monitoring and Prediction" is to Embedded Maturity-Strength Measurement System (WE-MSMS) by non-destructive approach is designed and developed using digital DS18B20 Temperature Sensor array probe, microcontroller, Zig-Bee network and IoT cloud.
- The other objective of the invention is to provide a Zig-Bee network with application software for maturity-strength measurement by non-destructive approach to estimate the concrete compressive strength of structure. A portable Wireless Temperature Sensor node with Zig-Bee network can be placed in the fresh concrete initially at construction site.
- The other objective of the invention is to provide a thereafter every 5-minute or according to time setting till 28-days, time-temperature updated information will recorded in the database using Zig-Bee network and IoT cloud.

- The other objective of the invention is to provide an eight-DS18B20 Temperature Sensors in an each sensor node. Four-Sensor Nodes Java platform based Application software is designed and used in the construction site's main server.
- The other objective of the invention is to provide a designed system useful for civil engineer to real-time prediction of in-situ concrete compressive strength at any-time and any-where with the help of wireless technologies and IoT cloud and also provide the real-time grading of concrete in various type of construction work.
- The other objective of the invention is to provide wireless and Real Time - Temperature updated recording and monitoring, and the computation of concrete strength by Maturity (TTF), equivalent age and temperature adjusted age functions.
- The other objective of the invention is to implement wireless maturity-strength measuring systems (WE-MSMS) on different construction sites and connected with IoT cloud for real-time prediction of in-situ concrete compressive strength.
- The other objective of the invention is to support ASTM C-1074 and IS 516:1959 Standards.

SUMMARY OF THE INVENTION

Existing System: CN865 Concrete Maturity Meter https://www.impact-test.co.uk/products/6581-6-channel-concrete-maturity-meter/	Our system: In-situ IoT cloud Based Concrete Maturity-strength Measurement Systems for Real Time Concrete strength Monitoring and Prediction
How it Works?	
The CN865 has 6 temperature channels and digitally displays the value of all channels.	It has 8 digital temperature probe with each Node
It can be connected directly to a PC or laptop via a USB interface for fast downloading of the readings.	Wireless connectivity via Zigbee network to a main construction site server.
Using the supplied Microsoft Excel™ templates, the Concrete Maturity is calculated from elapsed time versus temperature and is quoted as a "Maturity number".	Concrete Maturity calculation and prediction of concrete compressive strength using Excel.

The Maturity number can be used to determine when concrete has cured sufficiently for formwork to be removed.	Maturity and Compressive strength number can be used to remove the formwork.
In order to determine concrete maturity, sacrificial temperature measurement cables are 'cast into' the concrete and cut off at the surface when the concrete has cured.	Digital one-wire DS18B20 used to determine concrete maturity.
This Concrete Maturity Meter is a compact unit, making it ideal for site work in all conditions and climates.	Compact and wireless its ideal for site construction work in all conditions and climates.
The maturity meter is simple to use, finishing the job in the shortest possible time.	Wireless and in-situ
Features & benefits	
Logger for 6 x K or T type probes	32- Digital one-wire DS18B20 probe
Type K extension leads	8- Temperature sensors in Each node.
Software, suitable for Windows 2000, XP and Vista	Software Windows-7 and Windows-10
USB connecting cable	Wireless connectivity and IoT cloud
Our standard thermocouple wire & plugs may be used, for example TH088 & TH095.	Digital one-wire DS18B20 probe used.
Benefits with the implementation of Concrete Maturity Measurement Systems.	
Concrete Maturity Meter is a compact unit	Handy and compact
Concrete maturity measurement	in-situ concrete maturity and concrete strength measurements based on time-temperature history and age.
Support ASTM C-1074 & ASTM C-918	Support ASTM C-1074 and IS 516:1959 Standards.
Construction personnel with minimal training can easily implement monitoring function.	Same

DIAGRAM

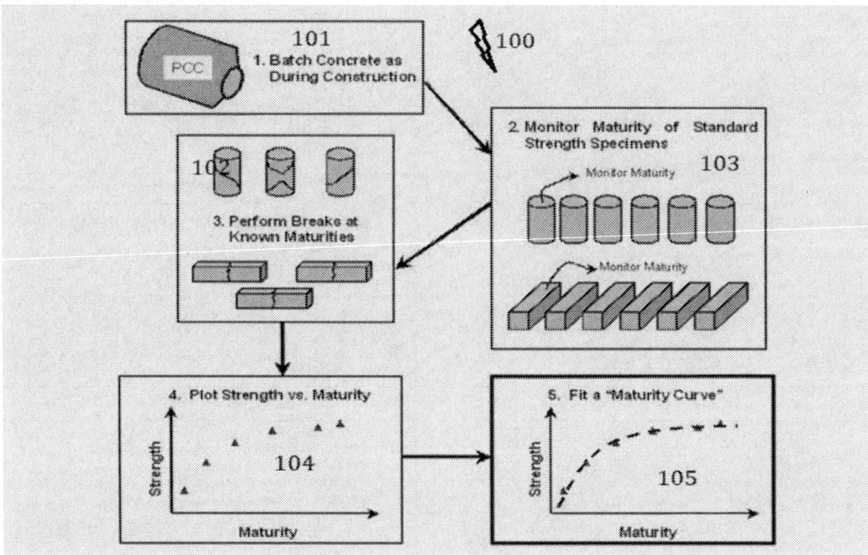

Fig.1: In-situ IoT cloud Based Concrete Maturity-Strength Measurement Systems flow chart (Prior Art Search)

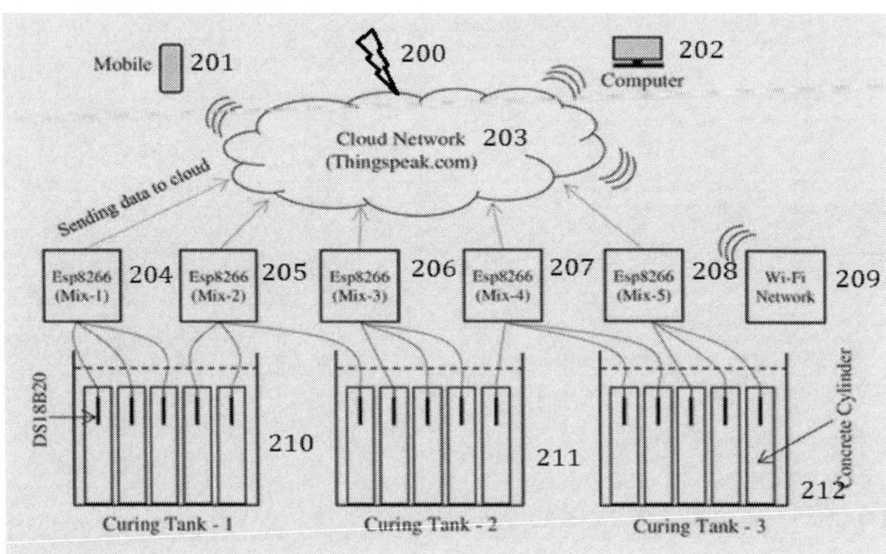

Fig. 2: In-situ IoT cloud Based Concrete Maturity-Strength Measurement Systems for Real Time Concrete strength Monitoring and Prediction (Prior Art Search)

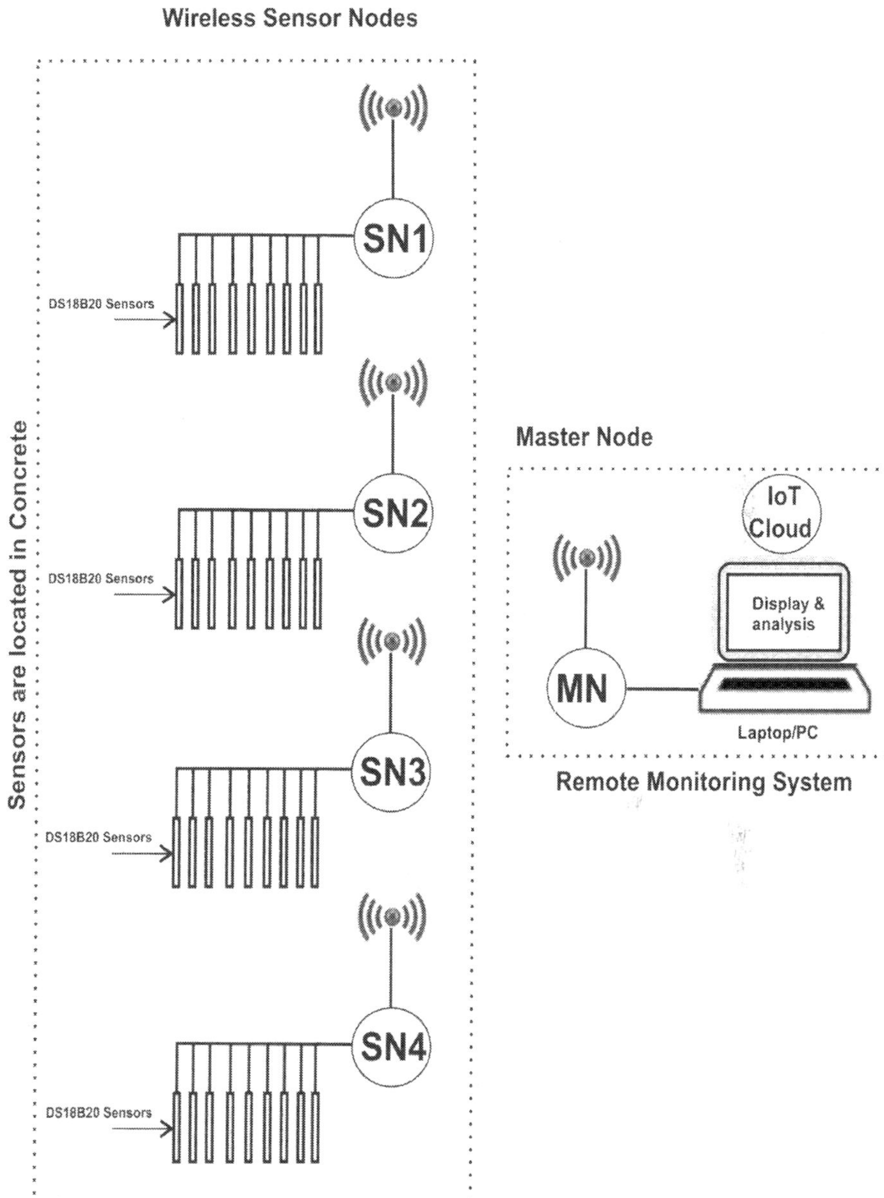

Fig. 3: Basic block of wireless sensor network for EW-MSMS

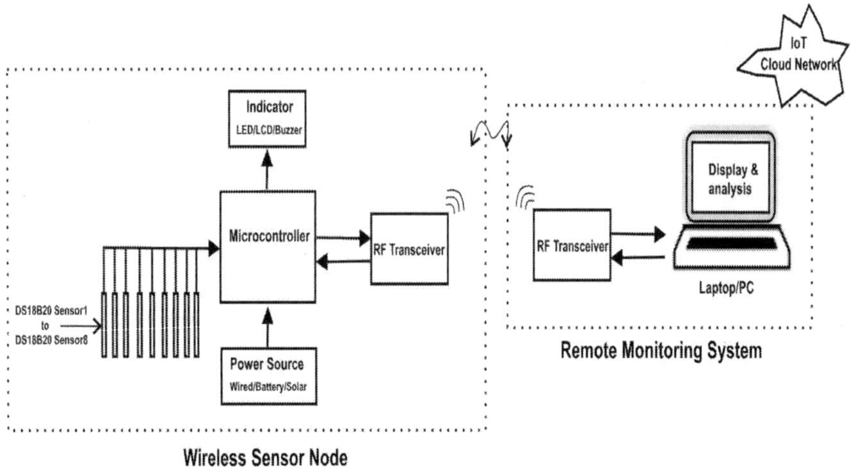

Fig. 4: Basic block of wireless sensor network for EW-MSMS

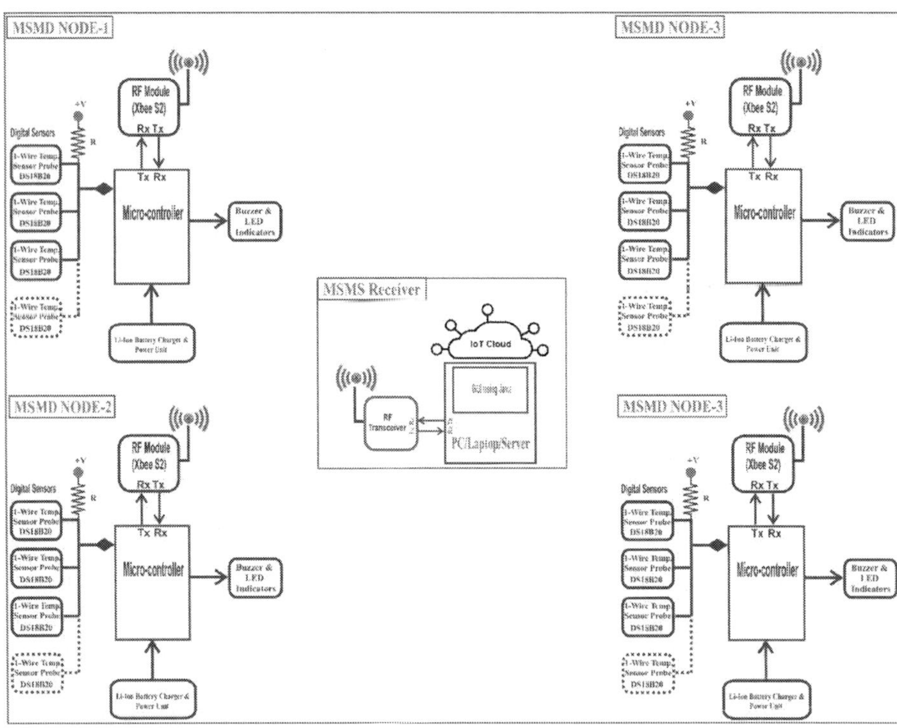

Fig. 5: Functional Block diagram of Wireless Embedded Maturity-Strength Measuring System

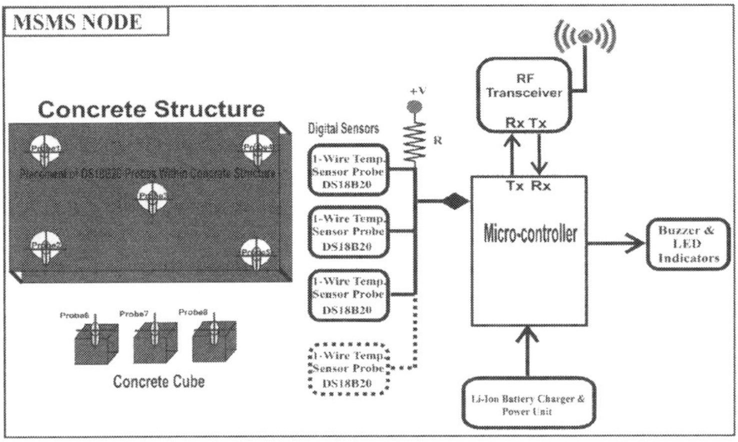

Fig. 6: Block Diagram of EW-MSMS Node.

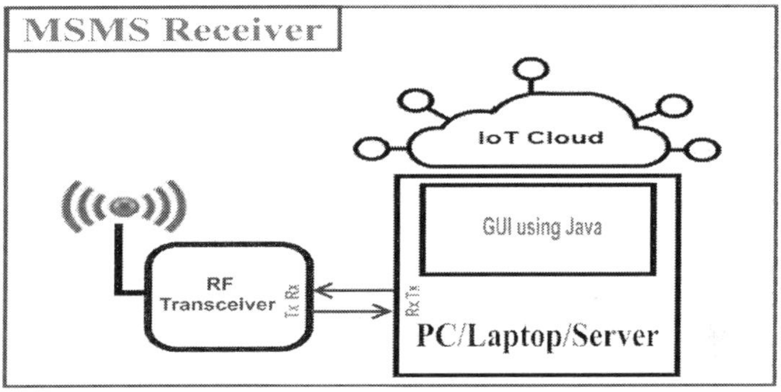

Fig. 7: Block Diagram of EW-MSMS Receiver

Fig. 8: Complete EW-MSMS Setup

DESCRIPTION OF THE INVENTION

Wireless Sensor Network (WSN) for Maturity-Strength Measuring System:

The designed wireless sensor network (WSN) is shown in figure 3. The network consists of sensor nodes (SNs) and master node (MN). The WSN consist of four wireless sensor nodes, namely, SN1, SN2, SN3 and SN4.

These SNs are located at different places in the construction field and all temperature sensors are embed in the concrete and sending temperature information to the master node (MN) for the estimation of in-place maturity-strength of the concrete. Then send the same information on cloud for access and further analysis.

The received data by SN1, SN2, SN3 and SN4 are displayed on the laptop to allow the user to monitor the concrete maturity and predict compressive strength and the same on cloud. MN is connected to the laptop, thus requiring no battery power source. The SN1, SN2, SN3 and SN4 are away from MN and their energy is derived from their individual power supplies. This means that SN1, SN2, SN3 and SN4 can be powered by the main AC source or battery or solar power source.

Wireless Sensor Node (SN) for Maturity-Strength Measuring System:

Figure 4: show the basic block of wireless sensor node (SN) for Maturity-strength measuring system. It consists of wireless sensor node and remote monitoring system.

Figure 4: Basic block of wireless sensor network for EW-MSMS

Each wireless sensor node is equipped with maximum eight numbers of 1-wire digital temperature sensors DS18B20, an ATMega 328P Microcontroller, Xbee transmitter, LED and buzzer indicator, and LCD display unit (optional). The entire sensor node (SN) is powered by the use of Li-ion battery backup power supply with charging circuitry arrangement. The sensor nodes (SNs) comprises of temperature sensors for measurement of temperature parameter of concrete structure and also concrete cube samples.

The remote monitoring system consist of Zig-Bee receiver in connection with personal computer or laptop or desktop or even to the raspberry Pi. The Graphical User interface is designed using JAVA language platform for monitoring and the measurement of thermal parameter and analysis of early age (fresh) concrete.

This arrangement is done for online real-time displaying temperature parameter recorded by each sensor embed in concrete cube which is under measurement. Hence the same arrangement is applied for finding the in-situ maturity-strength of concrete structure in field.

Wireless Embedded Maturity-Strength Measuring System (EW-MSMS)

The entire functional block diagram of wireless embedded Maturity-strength measuring system is as shown in figure 5. It consists of four sensor nodes (SNs) and remote monitoring system for data acquisition and thermal analysis of early age (fresh) concrete.

The entire Wireless Embedded Maturity-Strength Measuring System (EW-MSMS) is divided into five functional sections as follows,

Sensing temperature data from concrete structure /concrete mix design cubes by using multiple temperature sensors.

Data acquisition from each digital temperature sensor and to process on received data.

Data transmission using Zig-Bee wireless communication devices

Data reception using Zig-Bee wireless communication device.

Data analysis of received data by personal computer/Laptop.

The detail description of each functional block of Wireless Embedded Maturity-Strength Measuring System (EW-MSMS) as follows:

EW-MSMS Transmitter Node

Fig.6: The EW-MSMS measurement system's each data transmitting node consists of multiple numbers of 1-wire digital temperature sensor embed in concrete cube that are interfaced with an AT Mega 328PU eight bit microcontroller, a Xebec transmitter device for wireless data transmission, a Li-Ion battery unit for power backup with charging facility and LED and Buzzer indicator. In the present system there are 4 – nodes.

EW-MSMS Receiver Node

The EW-MSMS measurement system's data receiver section as shown in figure 7. consists of Xbee receiver connected to the personal computer terminal as shown in figure8. It is equipped with data acquisition and GUI software and data analysis software for the prediction of maturity strength of concrete cube under measurement.

Hardware Setup of Wireless Embedded MSMS

The complete hardware setup of Wireless embedded Maturity-Strength measurement system consisting Laptop, four Sensor Nodes (SNs), and remote receiver master node (MN) with IoT cloud is shown in figure 8.

Present Methods and Mathematical Models

Concrete is one of the most widely used in mixture form for any form of architectural structures, pavements, bridges, roads, runways, parking structures, dam and pipes etc. During initial state of concrete, in first few hours of consolidation if proper care is not taken, one can easily observe early cracks in solidified structure or leakages in structure and even sometimes entire structure can collapse that may turn into disastrous situation because of premature construction work.

Therefore, it is highly and most essential to measure and monitor the maturity strength of the concrete using in-place non-destructive technique is the only scientific way to prevent failure of the concrete structure. This is the main aspect in concrete structure development, the main emphasis is on strength measurement of the concrete structure.

There are also two major methods used for measuring the concrete strength, viz. destructive and non-destructive method. Destructive maturity strength measuring methods are time consuming as well as costly and are not suitable for the early age of concrete strength monitoring. These methods are not available at construction sites. Mostly, the concrete maturity was calculated using the Nurse-Saul and Arrhenius mathematical model/equations.

The present non-destructive methods are suitable for early ages of concrete strength monitoring. These are based on Thermocouples, Dallas Semiconductor iButton®, Nomadics, intelliRock™ and Identic Solutions i-Q Tag devices.

But there are some drawbacks of these systems such as bulky in size, direct wired measurement on site, limited channels, processing software limitations, non-continuous monitoring and costly.

The presented EW-MSMS is fully wireless technique based on the Nurse-Saul and Arrhenius mathematical model/equations. It minimizes drawbacks mentioned above. The key features of presented work are mentioned below.

- Multichannel facilities
- Wireless measurement on the site at anywhere and anytime

- Compact in size and Portable
- Software support (Java based GUI application)
- 24X7 Continuous monitoring using wireless techniques and IoT cloud
- Compact system setup
- Low Cost

Destructive method and non-destructive methods are used for measurement of concrete maturity. Destructive method was used for the actual physical measurement of strength of concrete by applying appropriate mechanical force on specimen structure. The performed during actual curing process of samples, it results in achieving compressive strength for sample under various environmental conditions.

The results obtained from non-destructive EW-MSMS correlated in concrete strength measurement process. The measured parameters from the distinct nodes are used for analysis and also prediction of Concrete Compressive Strength (CCS) using Raspberry Pi -3 based minicomputer or computer.

The compressive strength of concrete is measured at two places simultaneously i.e. in laboratory and on construction site. It is measured at ages of 1, 3, 7, 14 and 28 days in laboratory by using cube test and this data is stored in Micro SD card of the Raspberry Pi-3 minicomputer or computer. At the same time, it is measure on construction site and transfer and available the information on web using IoT cloud. The prediction of the compressive strength of the structure is represented in graphical format i.e. deviation of concrete compressive strength verses age using GUI designed with JAVA based software.

The wireless system is designed and implemented for monitoring, analysis and also prediction of compressive strength of the concrete which is useful to prevent architectural structure failure. In sensor nodes are designed to measure temperature development in early age concrete structure using ATmega328PU microcontroller based Arduino platform which transmits the sensed parameters in real time to the Raspberry Pi-3 minicomputer at receiving end.

The IoT cloud based Raspberry Pi-3 or minicomputer acts as receiver used for data analysis and maturity strength prediction purpose. Received data from Arduino platform transmitting nodes was also sent on data cloud using Internet platform anywhere using internet connectivity.

The designed system has four - wireless nodes. Each wireless node consists of eight (08) DS18B20 temperature sensors to measure temperature inside the concrete at construction site. Each wireless node transmits sensed temperature parameter to the base station using Zig-Bee wireless communication.

The base station consists of combination of internet, Zig-Bee and Raspberry Pi-3 minicomputer or computer. The received data is used for analysis and prediction of concrete strength using JAVA based Graphical User Interface (GUI).

Material and Method

The present system is divided into two parts; Transmitter Section using Zig-Bee and Arduino Platform and Receiver Section using Zig-Bee and Raspberry Pi Platform or IoT cloud based computer. The transmitter section consists of wireless nodes using Zig-Bee. In four nodes are used. Each node consists of following components.

Transmitter Section:

- *Temperature Sensors:* The DS18B20 is temperature sensing module which is used for temperature measurement. It is a small, 1-wire sensor digital sensing module with built in 12bit ADC. It gives temperature output in terms of degree Celsius or degree Fahrenheit. In each node, eight DS18B20 temperature sensors are immersed in concrete cube / structure.
- *Microcontroller:* The Atmega 328PU microcontroller is used as a part of sensor node. It is 28-pin, a single-chip microcontroller manufactured by Atmel. The output of each temperature sensor is connected to ATmega 328PU microcontroller using 1-wire interface. The ATmega 328PU microcontroller is easy to interface with other system using Arduino platform.
- *Zig-Bee Module:* The Xbee S2C is used for wireless data transmission. The measured temperature parameters in Celsius collected by ATmega328PU microcontroller are sent to the base station using Zig-Bee module. The Zig-Bee is configured as End device. The configuration of Xbee S2C has done through the X-CTU software by Digi International.

Receiver Section:

The receiver section also called as base station. It consists of Raspberry Pi-3 microcomputer or computer and data receiving Zig-Bee module. This section consists of following parts.

- *Zig-Bee Module:* It is at receiver section. This Zig-Bee is configured as coordinator. The data sent by transmitter section is collected at base station using Zigbee module. The configuration of Zig-Bee S2 has done through the X-CTU software by Digi International
- *Raspberry Pi and computer:* Raspberry Pi is 64-bit minicomputer with Broadcom BCM2837 and Quad Core Cortex A53. It work as mini computing device with Raspbian operating system. The data is stored in Raspberry Pi or computer for analysis.
- *Graphical User Interface (GUI):* The JAVA based GUI is developed for data analysis. The graphical analysis is done for temperature-time history. By using this graphical information, concrete strength can be predicated to avoid any structural failure due to premature work.

Wireless Communication

The data stored in Raspberry Pi or computer can be sent on cloud using Internet (IoT). Also this data and graphical information can be shared on email to user.

Embedded Wireless-Maturity-Strength Measurement System (EM-MSMS): EW-MSMS is a real-time, wireless communication based non-destructive way for continuous monitoring and analyzing for the in-place measurement of concrete compressive strength (CCS) at early ages. Destructive and Non-destructive tests are conducted for estimating compressive strength of concrete in laboratory (cube) and on the field (during construction).

The difference between two types of tests results was found in the range of 10% to 20% at approx. Thus Embedded Wireless-Maturity-Strength Measurement System (EW-MSMS) methodology has more potential to be used in-place compressive strength accurately for better results in concrete maturity under measurement.

Maturity-Strength Measuring System Hardware Design

The hardware details of the present EW-MSMS system which is useful for prediction of concrete compressive strength. Hardware design part includes Temperature sensor node (SN), wireless Transmitter unit and wireless Receiver Unit.

The wireless sensor network (WSN), wireless sensor node (SN) and various components used in WSN. It also covers details, features and application of selected devices such temperature sensors (DS18B20), microcontroller (ATMega328P) & Zig-Bee (Xbee-S2C) in the present system.

Maturity-Strength Measuring System Software

The software details of present works. Software tools includes WSN configuration with X-Bee S2c using X-CTU software platform, Sensor Node (SN) coding with ATMega328P microcontroller using Arduino Uno software platform and graphical user interface programming (GUI) using JAVA software platform .It also cover the development of data analysis forms for the presentation of measured data is discussed in detail.

Field Results

The about experimental work carried out on comparison results of concrete mix design cube test of M15, M20, & M25 samples using universal testing machine and discuss the results obtained from the above mentioned tests using wireless maturity-strength measuring system (WE-MSMS).

On basis of data collection made by present system the same utilized for measurement of compressive strength function or model suggested by Nurse-Saul Maturity Method, F. Hansen and Pedersen function and CEB-FIP model. It also discloses.

OUTCOME

Our Invention "In-situ IoT Cloud Based Concrete Maturity-strength Measurement Systems for Real Time Concrete strength Monitoring and Prediction" is to a Wireless Embedded Maturity-Strength Measurement System (WE-MSMS) is designed and developed using digital DS18B20 Temperature Sensor array probe, microcontroller and Zig-Bee network with application software for maturity-strength measurement by non-destructive approach to estimate the in-situ concrete compressive strength of structure. A portable Wireless Temperature Sensor node with Zig-Bee network can be placed at construction site in the fresh concrete initially. Four-Sensor Nodes GUI Java platform based Application software is designed and used in the main server. In each sensor node there are eight- digital DS18B20 Temperature Sensors. For the measurement DS18B20 Temperature Sensor array probe embed into the fresh concrete, each DS18B20 Temperature Sensor with unique code is recorded and maintain in the database at main server. Thereafter every 5-minute or according to time setting till 28-

days, time-temperature updated information will recorded in the central database using Zig-Bee network and IoT cloud. The designed system is useful for civil engineer to real-time prediction of in-situ concrete compressive strength at anytime and anywhere with the help of wireless technologies and IoT cloud which decide in-place grading of concrete in concrete construction work. Wireless maturity-strength measuring systems (WE-MSMS) can be implemented on different construction sites and connected with IoT cloud for real-time prediction of in-situ concrete compressive strength.

8. According to claim1# the invention is to an "In-situ IoT cloud Based Concrete Maturity-strength Measurement Systems for Real Time Concrete strength Monitoring and Prediction" is to a Wireless Embedded Maturity-Strength Measurement System (WE-MSMS) is designed and developed using digital DS18B20 Temperature Sensor array probe and microcontroller and also Zig-Bee network with application software for maturity-strength measurement by non-destructive approach to estimate the concrete compressive strength of structure.
9. According to claim1,2# the invention is to an portable Wireless Temperature Sensor node with Zig-Bee network can be placed at construction site in the fresh concrete initially. Four-Sensor Nodes GUI Java platform based Application software is designed and used in the main server.
10. According to claim1,2,3# the invention is to an portable Wireless Temperature sensor node there are eight- DS18B20 Temperature Sensors in array probe.
11. According to claim1,2,3,4# the invention is to the DS18B20 Temperature Sensor array probe embed into the fresh concrete, each DS18B20 Temperature Sensor with unique code is recorded and maintain in the database at main server.
12. According to claim1,2,3,4,5# the invention is to an 5-minute or according to time setting till 28-days, time-temperature updated information will recorded in the database using Zig-Bee network and IoT cloud.
13. According to claim1,2,4,5,6# the invention is to an designed system is useful for civil engineer to real-time prediction of in-situ concrete compressive strength at anytime and anywhere with the help of wireless technologies and IoT cloud and decide the grading of concrete in construction work .
14. According to claim1,2,4,5,6,7# the invention is to an portable Wireless Temperature Sensor node can be powered by the main

AC source or battery or solar power source for continuous system functioning.

15. According to claim1,2,4,5,6,7,8 # the invention is to an wireless maturity-strength measuring systems implemented on different construction sites and connected with IoT cloud for real-time prediction of in-situ concrete compressive strength.

REFERENCES

[1] Abhigna, B.S., Chakraborty, S., Kumar, V., Medha: A collaborative E-concept framework, Proceedings of the 2018 International Conference On Communication, Computing and Internet of Things, IC3IoT 2018, 2019, pp. 117–122, 8668127

[2] Abidin, S., Vadi, V.K., Rana, A., On Confidentiality, Integrity, Authenticity, and Freshness (CIAF) in WSN, Advances in Computer, Communication and Computational Sciences, 2020, pp 87-97

[3] ahay, A., Choudhury, T., Kumar, V., Hierarchical Evolution of Digital Games, Proceedings of 2019 International Conference on Computational Intelligence and Knowledge Economy, ICCIKE 2019, 2019, pp. 381–386, 9004378

[4] Area: Communication, Frequency modulation, Networks, wireless communication

[5] Area: Communication, Frequency modulation, Networks, wireless communication

[6] Area: Communication, Frequency modulation, Networks, wireless communication

[7] Area: Communication, Frequency modulation, Networks, wireless communication

[8] Arora, M., Kumar, V., Dave, M., Hybrid cloud infrastructure for load balancing and scaling with vm path optimization using ant colony optimization, Journal of Advanced Research in Dynamical and Control Systems, 2020, 12(5 Special Issue), pp. 660–674

[9] Arora, M., Kumar, V., Dave, M., Task scheduling in cloud infrastructure using optimization technique genetic algorithm, Proceedings of the World Conference on Smart Trends in Systems, Security and Sustainability, WS4 2020, 2020, pp. 788–793, 9210303

[10] Bansal, R., Gupta, A., Singh, R., Nassa, V.K., Role and impact of digital technologies in E-learning amidst COVID-19 pandemic, Proceedings - 2021 4th International Conference on Computational Intelligence and Communication Technologies, CCICT 2021, 2021, pp. 194–202

[11] Bansal, R., Obaid, A.J., Gupta, A., Singh, R., Pramanik, S., Impact of Big Data on Digital Transformation in 5G Era, Journal of Physics: Conference Seriesthis link is disabled, 2021, 1963(1), 012170

[12] Chauhan, S.S., Khare, A.K., Kinematic analysis of the ABB IRB 1520 industrial robot using roboanalyzer software, Evergreenthis link is disabled, 2020, 7(4), pp. 510–518

[13] Chauhan, S.S., Khare, A.K., Method of Obtaining Planar State of Stress Using Mohr's Circle—Some Typical Cases, Lecture Notes in Mechanical Engineering, 2022, pp. 33–38

[14] Goel, S., Tripathi, A., Choudhury, T., Kumar, V. Parkinson's Disease Detection using Soft Computing Technique, Proceedings of the 2019 8th International Conference on System Modeling and Advancement in Research Trends, SMART 2019, 2020, pp. 3–6, 9117384

[15] Govindarajan, V., Sivakumar, R., Patil, P.P., Kannan, M., Ramesh, B., Effect of Tungsten Carbide Addition on the Microstructure and Mechanical Behavior of Titanium Matrix Developed by Powder Metallurgy Route, Advances in Materials Science and Engineeringthis link is disabled, 2022, 2022, 2266951

[16] Gupta, A., Singh, R., Nassa, V.K., Sharma, P., Koti, K., Investigating Application and Challenges of Big Data Analytics with Clustering, 2021 International Conference on Advancements in Electrical, Electronics, Communication, Computing and Automation, ICAECA 2021, 2021

[17] Jain, P., Vijay, S., & Gupta, S. C. (2009). ATM Network QoS Parameters and Performance Analysis using OPNET Simulator. International Journal of Recent Trends in Engineering, 2(4), 228.

[18] Jain, P., Vijay, S., & Gupta, S. C. Performance Analysis & QoS Guarantee in ATM Networks. Global Journal of computer Science and Technology, 131-136.

[19] Jain, S., Choudhury, T., Kumar, V., Kumar, P., Detecting malware and analysing using sandbox evasion, Proceedings of the 2018 International Conference On Communication, Computing and Internet of Things, IC3IoT 2018, 2019, pp. 111–116, 8668149

[20] Kala, S., Nandan, H., Sharma, P., Shadow and weak gravitational lensing of a rotating regular black hole in a non-minimally coupled Einstein-Yang-Mills theory in the presence of plasma, European Physical Journal Plusthis link is disabled, 2022, 137(4), 457

[21] Kandpal, H., Kumar, A., Malik, A., Morphometric analysis and prioritization of sub-watersheds in Bino watershed, Uttarakhand: A remote sensing and GIS perspective, Indian Journal of Ecology, 2018, 45(3), pp. 475–478

[22] Kannan, M., Singh, S., Prasad, R.R., Synthetic Methods to Obtain Calcia-Stabilized Zirconia Powders—A Review, Lecture Notes in Mechanical Engineeringthis link is disabled, 2021, pp. 405–416

[23] Kukreti, A., Painolli, A.K., Rana, N., Where do we stand: factors affecting sustainable development, Procedia Environmental Science, Engineering and Management, Volume 8, Issue 2, 2021, Pages 583-589

[24] Kulkarni, S & Chima, P 2021 'Challenges faced by UK university students due to thcoronavirus crisis in the Higher Education' Preprints.org. https://doi.org/10.20944/preprints202102.0192.v1

[25] Kulkarni, S & Sodre, JR 2017, An Assessment of Lightweight Vehicles to Reduce Greenhouse Gas Emissions with Focus on Magnesium. in Thoughts and Refections on the use of Magnesium. Birmingham City University, pp. 84-91.

[26] Kulkarni, S, Chapman, C & Shah, H 2016 'Computational Fluid Dynamics (CFD) Mesh Independency Study of A Straight Blade Horizontal Axis Tidal Turbine' Preprints.org. https://doi.org/10.20944/preprints201608.0008.v1

[27] Kulkarni, S, Chapman, C, Shah, H & Edwards, D 2018, 'A computational design method for bio-mimicked horizontal axis tidal turbines', International Journal of Building Pathology and Adaptation, vol. 36, no. 2, pp. 188-209. https://doi.org/10.1108/IJBPA-06-2017-0029

[28] Kulkarni, S, Chapman, C, Shah, H, Parn, E & Edwards, DJ 2018, 'Designing an efficient tidal turbine blade through bio-mimicry: A systematic review', Journal of Engineering, Design and Technology, vol. 16, no. 1, pp. 101-124. https://doi.org/10.1108/JEDT-08-2017-0077

[29] Kulkarni, S, Chapman, C, Shah, H, Parn, E & Edwards, DJ 2018, 'Design Study of Horizontal Axis Tidal Turbine Blade', Mindanao Journal of Science and Technology. https://www.open-access.bcu.ac.uk/10089/

[30] Kulkarni, S, Edwards, DJ, Chapman, C, Hosseini, MR & Owusu-Manu, D-G 2019, 'A preliminary mechanical design evaluation of the Wikispeed car: for light-weighting implications', Journal of Engineering, Design and Technology, vol. 17, no. 1, pp. 230-249. https://doi.org/10.1108/JEDT-09-2018-0154

[31] Kulkarni, S, Edwards, DJ, Parn, E, Chapman, C, Aigbavboa, CO & Cornish, R 2018, 'Evaluation Of Vehicle Lightweighting To Reduce Greenhouse Gas Emissions With Focus On Magnesium Substitution', Journal of Engineering, Design and Technology, vol. 16, no. 6, pp. 869-888. https://doi.org/10.1108/JEDT-03-2018-0042

[32] Kulkarni, SS, Wang, L, Golsby, N & Lander, M 2021, 'Fluid-structure interaction based optimisation in tidal turbines: A perspective review', Journal of Ocean Engineering and Science. https://doi.org/10.1016/j.joes.2021.09.017

[33] Malik, A., Kumar, A., Kandpal, H., Morphometric analysis and prioritization of sub-watersheds in a hilly watershed using weighted sum approach, Arabian Journal of Geosciencesthis link is disabled, 2019, 12(4), 118

[34] Mohanakurup, V.K., Baghela, V.S., Kumar, S., Srivastava, P.K., Doohan, N.V., Soni, M., Awa, H., 5G Cognitive Radio Networks Using Reliable Hybrid Deep Learning Based on Spectrum Sensing, Green Internet of Things in 5G and Beyond, Hindawi, Volume 2022

[35] Mohta, M., Soni, N., Choudhury, T., Kumar, V., Market Analysis on Li-Fi Technology, Proceedings of the 2019 8th International Conference on System Modeling and Advancement in Research Trends, SMART 2019, 2020, pp. 139–143, 9117427

[36] Nagwan, H., Rana, N., Kumar, A., Tewari, P.C., Performance and emission characteristics of diesel engine using Karanja oil methyl ester and Eucalyptus oil fuel blends, Procedia Environmental Science, Engineering and Management, Volume 6, Issue 4, 2019, Pages 553-563

[37] Painoli, A.K., Bansal, R., Singh, R., Kukreti, A., Impact of digital marketing on the buying behavior of youth with special reference to Uttarakhand state, Big Data Analytics for Improved Accuracy, Efficiency, and Decision Making in Digital Marketing, 2021, pp. 162–182

[38] Parag, J. (2011). Hybrid mixed handover for call blocking probability analysis in wireless ATM network. Journal of Engineering and Computer Innovations, 2(4), 68-77.

[39] Sharma, C., Singh, M.P., Chaudhary, S., Cost Analysis of Solar-Wind Based Hybrid Renewable Energy System, Lecture Notes in Electrical Engineeringthis link is disabled, 2021, 700, pp. 2733–2746

[40] Sharma, S., Chandra, U., & Jain, P. (2017). A literature survey on automation of test data generation for branch coverage testing using genetic algorithm. International Journal of Computational Intelligence Research, 13(6), 1521-1531.

[41] Varshney, R., Jain, P., & Vijay, S. (2018, September). Massive MIMO Systems In Wireless Communication. In 2018 2nd International Conference on Micro-Electronics and Telecommunication Engineering (ICMETE) (pp. 39-44). IEEE.

Chapter - 10

Advanced Blockchain-based Distributed Using Machine Learning

Dr. Anil W. Kale

Mahatma Gandhi Mission's College of Engineering and Technology, India. Email: anil5474@gmail.com

Dr. Nandkishor P Karlekar

MGMCET Kamothe Navimumbai, India. Email: npkarlekar.mgmcet@gmail.com

Dr. Vaibhav Narawade

Ramrao Adik Institute of Technology, India. Email:vnarawade@gmail.com

Dr. Sanjay B. Waykar

Mahatma Gandhi Mission's College of Engineering and Technology, India. Email:sbwaykar@gmail.com

ABSTRACT

Our invention "Advanced Blockchain-based Distributed using Machine Learning." is a COVID-19 worldwide pestilence has modified our ways of life in various habits. It has made the need to continuously comprehend our position, which we've been found in collaboration with, as well as different points of interest like our body temp and to forestall the spread of such a viral disease. The above information is particularly significant for understudies, as their respect school will in general build their word related openness. To address this issue, a few control frameworks should be joined 100% of the time. People by and by utilize thermoelectric finders to look through an individual's temperature changes, card machines in transport courses to check teachers' presence, and web stages in which instructors can enlist the participation of the understudy for each class. People can utilize blockchain innovation to associate and interface and create an effective, clear COVID-19 global positioning framework. All through this paper, we begin discussing a new conveyed records detectability framework that ensures responsibility among the two understudies, schools, and the specialists to stay away from infection spread and help in collaboration following. Using blockchain innovation and homeroom Identity card, we support the execution of the old close by.

BACKGROUND

The major and developing utilization of virtual money has contributed essentially to the business' energetic conspicuousness. Frequently these individuals found out about blockchain and its applications due to Bitcoin.

Among the most engaging parts of conveyed record innovation are its outright clearness and freedom. It regularly spoke to huge quantities of individuals, especially while leading cash moves, since it didn't influence another person and was not checked by any single association. This is only one of the few applications for the bitcoin blockchain.

To our "new benchmark" because of COVID-19 and educational systems re-opening, blockchain innovations could work on the issue. Educational systems can utilize this self - observing apparatuses of who their understudies are meeting and to illuminate them assuming anyone shows indications, permitting them to actually manage what is going on.

Disseminated records are considered a substantially more straightforward and compelling strategy for redesigning datasets and EHRs (Electronic Health Records). Attempting to carry out blockchain innovation to sign patient data will give wiped out individuals, professionals, and some other workers with simple admittance to clinical records.

Administration clients regularly visit various treatment offices and professionals, so having the accessibility of data assembled in one area saves data and exertion. A key pair is utilized in this blockchain network, and the blockchain innovation is scattered among reliable individuals.

The decentralized resources are a truly important element of simply utilizing dispersed records innovation. The blockchain gives decentralized applications, which appear to be carefully marked kinds of agreements. Various agreements in wellbeing inclusion should be avowed and approached with deference, including the enrollment understanding, the doctor relationship arrangement, and the summing up agreement.

All of this is recorded in the blockchain. This outcomes in a more quiet focused system. Conveyed records are by and by utilized for sharing information, security frameworks, and wellbeing data, in spite of the fact that there is huge potential for development in parts of production network and medication recommending key preparation. Blockchain innovation has a lot of potential inside the schooling field too. Current methods, for example, Boards are mindful, virtual cash, and the Network.

For instance (however Cryptocurrency actually requires working frameworks to develop decentralized applications and associate straightforwardly to data sets) were utilized to endlessly get recognitions, acknowledge cash from colleagues, as well as give understudy cash as coupons through blockchain innovation.

Both these potential employments of blockchain-based that are right now being scrutinized in this ground likewise use blockchain to affirm multi-step confirmation, to empower account opening affirmation and send, to record copyrights, and considerably more. Attempting to move our concentrate away from blockchain, it likewise is beneficial to explore the cycles that have as of late arisen because of the avian flu influenza plague, as well as the possibility of small children applying to school in such circumstances.

Information about a kid's conduct and temp are becoming basic to guaranteeing the security and very well, all things considered. Three frameworks are especially significant for giving these subtleties: thermoelectric scanners, transport swipe cards, and electronic cooperation logbooks.

Infrared imaging distinguishes infrared photons and afterward utilizes the hotness energy got from them and afterward creates pictures that address the singular's temp. The hotter the element, the more clear the image. Thermoelectric cameras could be utilized to decide a singular's thermometer to inside 0.1 Fahrenheit.

OBJECTIVES

- The objective of the invention is to provide a "A Robust Framework for Governing Blockchain-based Distributed Ledgers during Covid-19 for Academic Establishments" is a COVID-19 worldwide pestilence has modified our ways of life in various habits.

- The other objective of the invention is to provide a made the need to continuously comprehend our position, which we've been found in collaboration with, as well as different points of interest like our body temp and to forestall the spread of such a viral disease.
- The other objective of the invention is to provide a above information is particularly significant for understudies, as their respect school will in general build their word related openness. To address this issue, a few control frameworks should be joined 100% of the time.
- The other objective of the invention is to provide a utilize thermoelectric finders to look through an individual's temperature changes, card machines in transport courses to check teachers' presence, and web stages in which instructors can enlist the participation of the understudy for each class.
- The other objective of the invention is to provide a People can utilize blockchain innovation to associate and interface and create an effective, clear COVID-19 global positioning framework. All through this paper, we begin discussing a new conveyed records detectability framework that ensures responsibility among the two understudies, schools.
- The other objective of the invention is to provide a specialists to stay away from infection spread and help in collaboration following. Using blockchain innovation and homeroom Identity card, we support the execution of the old close by.

SUMMARY

As expressed already, Blockchain innovation has an assortment of utilizations in training and medical services. In any case, seasonal influenza pestilence has made it conceivable to associate the over two factors in a brand different way: need of the moment is utilize circulated records innovations to help connection attempting to follow between many school-matured little youngsters, i.e., perceive people who might need to show up into contact with the contaminated conventional individual and assemble data regarding a portion of these contact data.

We recommend the improvement of a COVID-19 Tracking framework, which will consolidate the referenced beforehand processes using blockchain innovation to fabricate a framework wherein a youngster's support and temp will be reported and accessible to the junior college, the specialists, and the individual and their relatives.

The youngster should download and introduce the COVID Tracking framework for simple openness, as well as proposition up distinguishing proof data that should be validated. This remembered the pertinent information for their Reference Number (public identifier) and instruction ID card (school identifier) so the two partners can check the singular's very own data and since the two players ought to approach this global positioning framework.

After the youngster has without a doubt been confirmed as a local area part and selected as an understudy at a particular school, both these subtleties fundamental to the rudimentary, like class, fragment, and public.

DIAGRAM

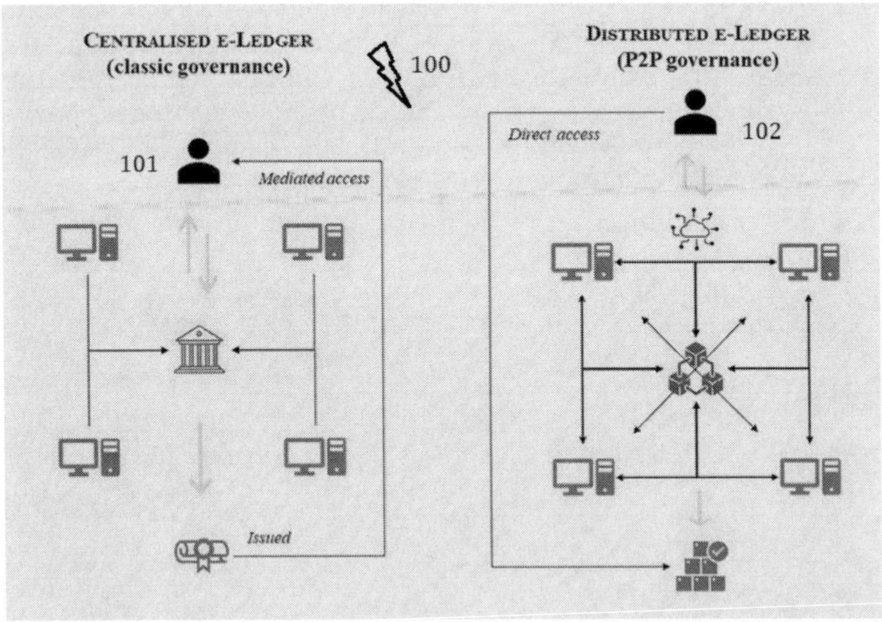

Fig. 1: A Robust Framework for Governing Blockchain-based Distributed Ledgers during Covid-19 for Academic Establishments Flow Chart.

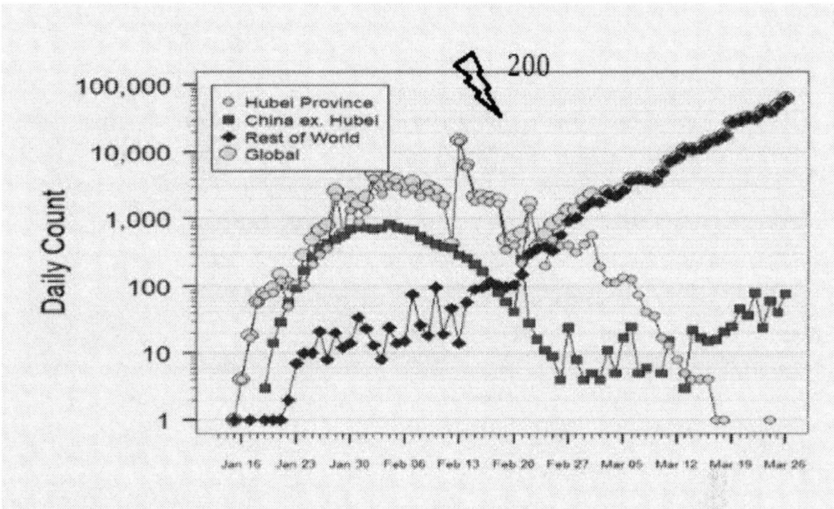

Fig. 2: A Robust Framework Graph

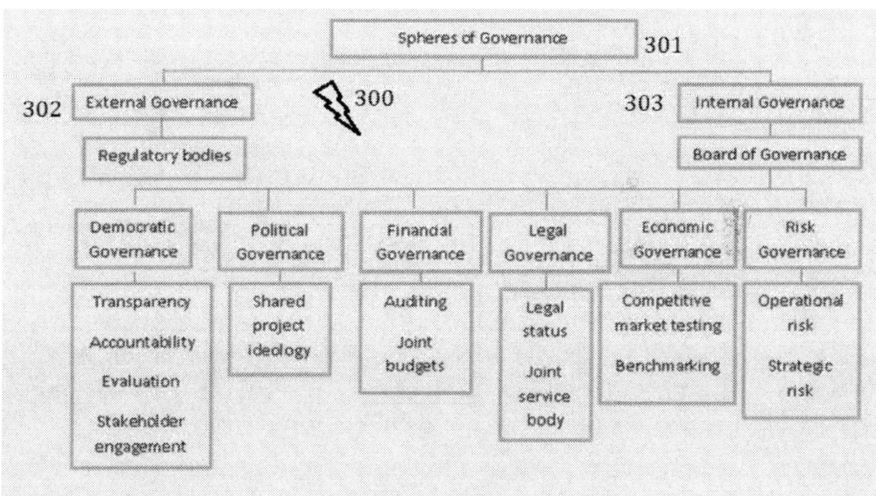

Fig. 3: Robust Framework for Governing

DESCRIPTION

Whenever a youngster boarded the transport, one's presence was recorded in the blockchain innovation through the card machine that they were utilizing to assess their schooling Identification card.

Whenever understudies come to school, their temperature was recorded, and this information is likewise added to the chain by means of the thermoelectric scanner.

During the day, every educator should accept understudy's participation existing for each time period and logbooks it in to no one's surprise, and that accordingly this information likewise is being acquainted with the square chain innovation and associated with each researcher in such a manner so every understudy's support can be gotten to freely during the day.

There at the last snapshot of every day, and encompassing temperature perusing was recorded, as well as the patient's presence is archived on the last transport ride home. The above information could likewise be approved by the framework to guarantee its accuracy prior to being shared on the web. Up-and-comers can actually take a look at their data and request a change assuming they have erroneous investment.

These equivalent advantages about utilizing a blockchain network incorporate the way that this is a really protected strategy that can be fiddled with, that purchasers have one of a kind identifiers and are certify, and it offers huge measures of promptly accessible data, which is particularly significant in a real crisis including a COVID specific occurrence.

On the off chance that a youngster examinations positive for such popular contamination, their classmates and everybody with whom the person would have come in contact can be without any problem.

All information is added to the blockchain by means of installment frameworks, endeavoring to make it protected and individual, even as blockchain network ensures that different understudies can't get to another patient's information, which is fundamental to keeping up with wellbeing and forestalling boundless frenzy and disarray in the event of a Covid flare-up at school.

Besides, the decentralized framework guarantees that It isn't just significant for instructive specialists to approach this information, and it is additionally critical to remember that now the current system will for sure require this information for genuine information and checking in such a sort of circumstance, and accordingly the decentralization would further develop productivity not just inside partners as well as between parties.

The structure is additionally scattered, which disposes of the chance of the information group losing. Generally speaking, blockchain supports the joining of significant frameworks to wind up making a viable record strategy that is promptly accessible to all specialists and people, speeding up in the methodology, instead of record-continuing to become kept offsite and simply accessible to a couple of put down accounts.

The world we are living in is verifiably difficult, and in this way we should continue mindfully. Indeed, even as "new reality" has become recognized, educational systems and scholastic organizations have started to jump up around the world. Our proposed COVID19 reporting framework could turn into an extremely effective approach to limiting and controlling this situation.

This method can give more precise and schools of assessing the quantity of patients and gives significant information which can be utilized to decide how we could best deal with a circumstance. As this is probably going to endure, frameworks should be set up to safeguard understudies, teachers, school work force, drivers, and relatives.

There appear to be countless gatherings in the present situation, which guarantees their wellbeing to some huge broaden. This should begin opening a few generally unseen universes of blockchain innovation, helping further in the innovation mix into our regular schedules.

OUTCOME

Our invention "A Robust Framework for Governing Blockchain-based Distributed Ledgers during Covid-19 for Academic Establishments" is a COVID-19 worldwide pestilence has modified our ways of life in various habits. It has made the need to continuously comprehend our position, which we've been found in collaboration with, as well as different points of interest like our body temp and to forestall the spread of such a viral disease. The above information is particularly significant for understudies, as their respect school will in general build their word related openness. To address this issue, a few control frameworks should be joined 100% of the time. People by and by utilize thermoelectric finders to look through an individual's temperature changes, card machines in transport courses to check teachers' presence, and web stages in which instructors can enlist the participation of the understudy for each class. People can utilize blockchain innovation to associate and interface and create an effective, clear COVID-19 global positioning framework. All through this paper, we begin discussing a new conveyed records detectability framework that ensures responsibility among the two understudies, schools, and the specialists to stay away from infection spread and help in collaboration following. Using blockchain innovation and homeroom Identity card, we support the execution of the old close by.

According to claim1# the invention is to a "A Robust Framework for Governing Blockchain-based Distributed Ledgers during Covid-19 for Academic Establishments" is a COVID-19 worldwide pestilence has modified our ways of life in various habits.

1. According to claim1# the invention is to a made the need to continuously comprehend our position, which we've been found in collaboration with, as well as different points of interest like our body temp and to forestall the spread of such a viral disease.
2. According to claim1,2,3# the invention is to a above information is particularly significant for understudies, as their respect school will in general build their word related openness. To address this issue, a few control frameworks should be joined 100% of the time.
3. According to claim1,2,3# the invention is to a utilize thermoelectric finders to look through an individual's temperature changes, card machines in transport courses to check teachers' presence, and web stages in which instructors can enlist the participation of the understudy for each class.
4. According to claim1,2,3,5# the invention is to a People can utilize blockchain innovation to associate and interface and create an effective, clear COVID-19 global positioning framework. All through this paper, we begin discussing a new conveyed records detectability framework that ensures responsibility among the two understudies, schools.
5. According to claim1,2,3,4# the invention is to a specialists to stay away from infection spread and help in collaboration following. Using blockchain innovation and homeroom Identity card, we support the execution of the old close by.

REFERENCES

[1] Abhigna, B.S., Chakraborty, S., Kumar, V., Medha: A collaborative E-concept framework, Proceedings of the 2018 International Conference On Communication, Computing and Internet of Things, IC3IoT 2018, 2019, pp. 117–122, 8668127

[2] Abidin, S., Vadi, V.K., Rana, A., On Confidentiality, Integrity, Authenticity, and Freshness (CIAF) in WSN, Advances in Computer, Communication and Computational Sciences, 2020, pp 87-97

[3] Ahay, A., Choudhury, T., Kumar, V., Hierarchical Evolution of Digital Games, Proceedings of 2019 International Conference on Computational Intelligence and Knowledge Economy, ICCIKE 2019, 2019, pp. 381–386, 9004378

[4] Area: Communication, Frequency modulation, Networks, wireless communication

[5] Area: Communication, Frequency modulation, Networks, wireless communication

[6] Area: Communication, Frequency modulation, Networks, wireless communication

[7] Area: Communication, Frequency modulation, Networks, wireless communication

[8] Arora, M., Kumar, V., Dave, M., Hybrid cloud infrastructure for load balancing and scaling with vm path optimization using ant colony optimization, Journal of Advanced Research in Dynamical and Control Systems, 2020, 12(5 Special Issue), pp. 660–674

[9] Arora, M., Kumar, V., Dave, M., Task scheduling in cloud infrastructure using optimization technique genetic algorithm, Proceedings of the World Conference on Smart Trends in Systems, Security and Sustainability, WS4 2020, 2020, pp. 788–793, 9210303

[10] Bansal, R., Gupta, A., Singh, R., Nassa, V.K., Role and impact of digital technologies in E-learning amidst COVID-19 pandemic, Proceedings - 2021 4th International Conference on Computational Intelligence and Communication Technologies, CCICT 2021, 2021, pp. 194–202

[11] Bansal, R., Obaid, A.J., Gupta, A., Singh, R., Pramanik, S., Impact of Big Data on Digital Transformation in 5G Era, Journal of Physics: Conference Seriesthis link is disabled, 2021, 1963(1), 012170

[12] Chauhan, S.S., Khare, A.K., Kinematic analysis of the ABB IRB 1520 industrial robot using roboanalyzer software, Evergreenthis link is disabled, 2020, 7(4), pp. 510–518

[13] Chauhan, S.S., Khare, A.K., Method of Obtaining Planar State of Stress Using Mohr's Circle—Some Typical Cases, Lecture Notes in Mechanical Engineering, 2022, pp. 33–38

[14] Goel, S., Tripathi, A., Choudhury, T., Kumar, V. Parkinson's Disease Detection using Soft Computing Technique, Proceedings of the 2019 8th International Conference on System Modeling and Advancement in Research Trends, SMART 2019, 2020, pp. 3–6, 9117384

[15] Govindarajan, V., Sivakumar, R., Patil, P.P., Kannan, M., Ramesh, B., Effect of Tungsten Carbide Addition on the Microstructure and Mechanical Behavior of Titanium Matrix Developed by Powder Metallurgy Route, Advances in Materials Science and Engineeringthis link is disabled, 2022, 2022, 2266951

[16] Gupta, A., Singh, R., Nassa, V.K., Sharma, P., Koti, K., Investigating Application and Challenges of Big Data Analytics with Clustering, 2021 International Conference on Advancements in Electrical, Electronics, Communication, Computing and Automation, ICAECA 2021, 2021

[17] Jain, P., Vijay, S., & Gupta, S. C. (2009). ATM Network QoS Parameters and Performance Analysis using OPNET Simulator. International Journal of Recent Trends in Engineering, 2(4), 228.

[18] Jain, P., Vijay, S., & Gupta, S. C. Performance Analysis & QoS Guarantee in ATM Networks. Global Journal of computer Science and Technology, 131-136.

[19] Jain, S., Choudhury, T., Kumar, V., Kumar, P., Detecting malware and analysing using sandbox evasion, Proceedings of the 2018 International Conference On Communication, Computing and Internet of Things, IC3IoT 2018, 2019, pp. 111–116, 8668149

[20] Kala, S., Nandan, H., Sharma, P., Shadow and weak gravitational lensing of a rotating regular black hole in a non-minimally coupled Einstein-Yang-Mills theory in the presence of plasma, European Physical Journal Plusthis link is disabled, 2022, 137(4), 457

[21] Kandpal, H., Kumar, A., Malik, A., Morphometric analysis and prioritization of sub-watersheds in Bino watershed, Uttarakhand: A remote sensing and GIS perspective, Indian Journal of Ecology, 2018, 45(3), pp. 475–478

[22] Kannan, M., Singh, S., Prasad, R.R., Synthetic Methods to Obtain Calcia-Stabilized Zirconia Powders—A Review, Lecture Notes in Mechanical Engineeringthis link is disabled, 2021, pp. 405–416

[23] Kukreti, A., Painolli, A.K., Rana, N., Where do we stand: factors affecting sustainable development, Procedia Environmental Science, Engineering and Management, Volume 8, Issue 2, 2021, Pages 583-589

[24] Kulkarni, S & Chima, P 2021 'Challenges faced by UK university students due to thcoronavirus crisis in the Higher Education' Preprints.org. https://doi.org/10.20944/preprints202102.0192.v1

[25] Kulkarni, S & Sodre, JR 2017, An Assessment of Lightweight Vehicles to Reduce Greenhouse Gas Emissions with Focus on Magnesium. in Thoughts and Refections on the use of Magnesium. Birmingham City University, pp. 84-91.

[26] Kulkarni, S, Chapman, C & Shah, H 2016 'Computational Fluid Dynamics (CFD) Mesh Independency Study of A Straight Blade Horizontal Axis Tidal Turbine' Preprints.org. https://doi.org/10.20944/preprints201608.0008.v1

[27] Kulkarni, S, Chapman, C, Shah, H & Edwards, D 2018, 'A computational design method for bio-mimicked horizontal axis tidal turbines', International Journal of Building Pathology and Adaptation, vol. 36, no. 2, pp. 188-209. https://doi.org/10.1108/IJBPA-06-2017-0029

[28] Kulkarni, S, Chapman, C, Shah, H, Parn, E & Edwards, DJ 2018, 'Designing an efficient tidal turbine blade through bio-mimicry: A systematic review', Journal of Engineering, Design and Technology, vol. 16, no. 1, pp. 101-124. https://doi.org/10.1108/JEDT-08-2017-0077

[29] Kulkarni, S, Chapman, C, Shah, H, Parn, E & Edwards, DJ 2018, 'Design Study of Horizontal Axis Tidal Turbine Blade', Mindanao Journal of Science and Technology. https://www.open-access.bcu.ac.uk/10089/

[30] Kulkarni, S, Edwards, DJ, Chapman, C, Hosseini, MR & Owusu-Manu, D-G 2019, 'A preliminary mechanical design evaluation of the Wikispeed

car: for light-weighting implications', Journal of Engineering, Design and Technology, vol. 17, no. 1, pp. 230-249. https://doi.org/10.1108/JEDT-09-2018-0154

[31] Kulkarni, S, Edwards, DJ, Parn, E, Chapman, C, Aigbavboa, CO & Cornish, R 2018, 'Evaluation Of Vehicle Lightweighting To Reduce Greenhouse Gas Emissions With Focus On Magnesium Substitution', Journal of Engineering, Design and Technology, vol. 16, no. 6, pp. 869-888. https://doi.org/10.1108/JEDT-03-2018-0042

[32] Kulkarni, SS, Wang, L, Golsby, N & Lander, M 2021, 'Fluid-structure interaction based optimisation in tidal turbines: A perspective review', Journal of Ocean Engineering and Science. https://doi.org/10.1016/j.joes.2021.09.017

[33] Malik, A., Kumar, A., Kandpal, H., Morphometric analysis and prioritization of sub-watersheds in a hilly watershed using weighted sum approach, Arabian Journal of Geosciencesthis link is disabled, 2019, 12(4), 118

[34] Mohanakurup, V.K., Baghela, V.S., Kumar, S., Srivastava, P.K., Doohan, N.V., Soni, M., Awa, H., 5G Cognitive Radio Networks Using Reliable Hybrid Deep Learning Based on Spectrum Sensing, Green Internet of Things in 5G and Beyond, Hindawi, Volume 2022

[35] Mohta, M., Soni, N., Choudhury, T., Kumar, V., Market Analysis on Li-Fi Technology, Proceedings of the 2019 8th International Conference on System Modeling and Advancement in Research Trends, SMART 2019, 2020, pp. 139–143, 9117427

[36] Nagwan, H., Rana, N., Kumar, A., Tewari, P.C., Performance and emission characteristics of diesel engine using Karanja oil methyl ester and Eucalyptus oil fuel blends, Procedia Environmental Science, Engineering and Management, Volume 6, Issue 4, 2019, Pages 553-563

[37] Painoli, A.K., Bansal, R., Singh, R., Kukreti, A., Impact of digital marketing on the buying behavior of youth with special reference to Uttarakhand state, Big Data Analytics for Improved Accuracy, Efficiency, and Decision Making in Digital Marketing, 2021, pp. 162–182

[38] Parag, J. (2011). Hybrid mixed handover for call blocking probability analysis in wireless ATM network. Journal of Engineering and Computer Innovations, 2(4), 68-77.

[39] Sharma, C., Singh, M.P., Chaudhary, S., Cost Analysis of Solar-Wind Based Hybrid Renewable Energy System, Lecture Notes in Electrical Engineeringthis link is disabled, 2021, 700, pp. 2733–2746

[40] Sharma, S., Chandra, U., & Jain, P. (2017). A literature survey on automation of test data generation for branch coverage testing using genetic algorithm. International Journal of Computational Intelligence Research, 13(6), 1521-1531.

[41] Varshney, R., Jain, P., & Vijay, S. (2018, September). Massive MIMO Systems In Wireless Communication. In 2018 2nd International Conference on Micro-Electronics and Telecommunication Engineering (ICMETE) (pp. 39-44). IEEE.

CHAPTER - 11

Machine Learning and AI (Object Detection)

Dr. Sanjay B. Waykar

Mahatma Gandhi Mission's College of Engineering and Technology, India. Email:sbwaykar@gmail.com

Dr. Anil W. Kale

Mahatma Gandhi Mission's College of Engineering and Technology, India. Email: anil5474@gmail.com

Dr. Nandkishor P Karlekar

MGMCET Kamothe Navimumbai, India. Email: npkarlekar.mgmcet@gmail.com

Dr. Vaibhav Narawade

Ramrao Adik Institute of Technology, India. Email:vnarawade@gmail.com

ABSTRACT

Our Invention "AI AND Machine LEARNING. "is an automated airborne vehicle (UAV) can incorporate more than one cameras for catching picture information inside a field of view that depends partially upon the area and direction of the UAV under profound learning method. The development likewise a piece of the picture information can be handled and examination on the UAV to find objects of interest like moving articles, individuals, leaving things or vehicles, and utilize that data to figure out where to fly the robot to catch better picture information of those or other such items. The imagined innovation additionally distinguished the objects of interest can be tallied and the thickness, development, area, and conduct of those articles recognized and furthermore this can assist with deciding events, for example, gridlock or strange examples of passer-by development, just as to find people, flames, or other such items. The creation likewise be dissected by a distant framework or administration that has extra assets to give more exact outcomes approx. 98%.

BACKGROUND

As automated ethereal vehicles (UAVs) or "robots" become all the more innovatively progressed, and as the passage level expenses for these vehicles diminishes, there is interest in deciding approaches to use these gadgets to tackle different issues or perform different errands. Robots can work in areas and catch picture sees that would be hard to in any case get utilizing traditional cameras or gadgets. The capacity to use these picture sees is restricted, be that as it may, because of elements like the restricted handling limit, information stockpiling, and correspondences transfer speed of these gadgets. While bigger robots can be utilized that can uphold extra assets, these robots are altogether costlier to work and are limited with respect to where and how they can work.

Human-PC cooperation innovation be joined by the introduction of PC and produce and with PC programming and equipment advancement and by Gradually create, the presence of new method continually improves on the stream cycle of man-machine collaboration. As of late, with the presence of computerized reasoning innovation and improvement, the ceaseless reformist and advancement of related programming and equipment innovation, how to understand that more helpful man-machine collaboration becomes study area of interest, Various novel human-machine connection advances are kept on bringing out. Simultaneously, the ascent of important enterprises, for example, modest SUAV (UAV) with Popularization, in the dire need to a few, all the more effectively individuals diminishes the edge that automated plane is worked with the intelligent controlling method of automated plane so that Unmanned plane is progressively generally applied.

OBJECTIVES

- The objective of the invention is to a unmanned aerial vehicle (UAV) can include more than one cameras for capturing image data within a field of view that depends in part upon the location and orientation of the UAV under deep learning technique.
- The objective of the invention is to a portion of the image data can be processed and analysis on the UAV to locate objects of interest such as moving objects, people, leaving things or cars, and use that information to determine where to fly the drone

in order to capture higher quality image data of those or other such objects.

- The objective of the invention is to a identified the objects of interest can be counted and the density, movement, location, and behavior of those objects identified and also this can help to determine occurrences such as traffic congestion or unusual patterns of pedestrian movement, as well as to locate persons, fires, or other such objects.
- The objective of the invention is to the invention also be analyzed by a remote system or service that has additional resources to provide more accurate results approx. 98%.

SUMMARY

The framework via convey the automated aeronautical vehicle foundation of topographical position obtaining module, implanted picture handling stage, visual head, 4, face station part is established. Automated flying vehicle stage is multi-rotor automated flying vehicle, and automated plane is situated by geological position securing module, flight control Device can handle automated plane in open air independent drifting. Installed picture preparing stage and visual head, the framework are continued automated plane In visual head get for ensuing treatment superior quality picture.

Installed stage is the stage that can give enough functional abilities to picture preparing with realistic cycle unit (GPU). This stage is liable for the securing system and activity acknowledgment of picture, in down to earth application, can be filled in as by superior portable terminal overseer, the framework associates flight regulator by interface, and accordingly stage fills in as the correspondence on automated plane and ground. Stage conveys working framework Operation handling schedule. Earth station is answerable for observing the condition of four-pivot airplane, for indicating avigator and actually looking at the aftereffect of ongoing activity, Can be filled in as by note pad or keen terminal.

Activity Recognition Framework

Activity acknowledgment structure principally incorporates Video pre-separating, the guidance for producing shading surface figure and convolutional neural organizations model Practice and order.

1. as indicated by the one-see picture of automated plane pass back, automated plane avigator is chosen, strategy is utilizing

mouse or contact Screen, body (chest area) locale of avigator is irised out to come ;

2. chose locale is ventured into outline as per individual proportion and chooses the entire body area of individuals, while from one more In the center of perspective, the individual district of separate areas is chosen, and extricated outline by outline in the video succession utilizing track calculation Go out avigator locale and avigator is followed, and as indicated by the situation of setting locales, by entire human body foundation Region picture division out ;
3. stereo coordinating, as per the Stereo Matching Algorithm dependent on Block-coordinating, by the picture between the left and right perspective for parting Matched, the stereopsises figure in the wake of being parted, the personage involving avigator and foundation in divergence map ;
4. divergence map for acquiring is standardized. What's more, use edge esteem clearing out foundation. Acquire clean person image ;
5. character picture of contiguous two edge is done into distinction, acquires contrast picture succession.
6. as per the image grouping for creating, progressively the distinction picture utilizing various shadings delegate not in a similar time, is encoded, The person picture that the image of around 2s or somewhere in the vicinity is delivered is superimposed as into shading surface figure. Utilizing the strategy for sliding window, take each 5 casings The image of around 2s or so ;
7. assemble considerable measures of, the shading surface figure of chosen activity is done under different conditions to nonpartisan net by various working individual It is prepared. Preparing is done in specific reason workstation. After the finish of preparing, preparing boundary is transferred to into implanted picture handling stage.
8. on installed picture handling stage, utilizing the boundary for preparing to Real-time Collection and create shading surface Figure is ordered.
9. on installed picture handling stage, following, division, Stereo coordinating requests shading surface figure, and characterization exists individually Different strings is completed all the while, in order to the removal capacity for considerably

using processor. In the interim, by about the working out of picture, benefit Accelerated with realistic cycle unit so that preparing speed meets the necessity of constant.

Advantages of the Present Invention and Beneficial Effect

First, convolutional neural organizations are relocated to and are arranged with the implanted stage with committed designs processors (GPU) by the current creation On, sped up to speed up the speed of administration by equal calculation. second, the current development removes administrator area utilizing objective following algorism from video succession, proficiently takes care of nothing The issues, for example, man-machine in-flight camera float and complex foundation are upset, while lessening operand. The strategy and technique for separating stage Relatively, with working reach it is wide, exactness rate is high the qualities. third, the current development will be sped up with regards to the computing of picture utilizing realistic interaction unit so that preparing speed meets constant Requirement.

DIAGRAM

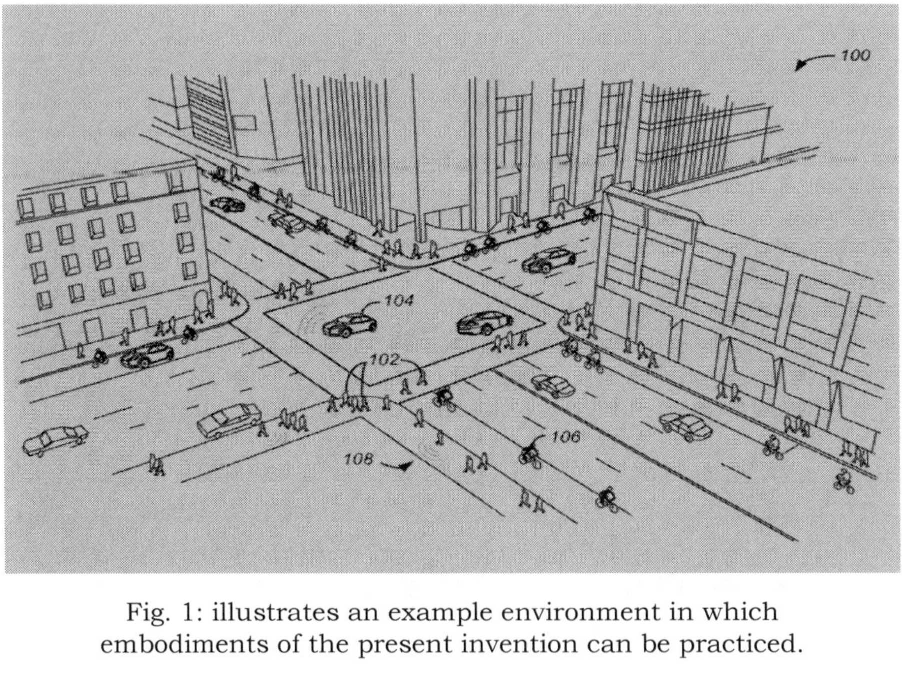

Fig. 1: illustrates an example environment in which embodiments of the present invention can be practiced.

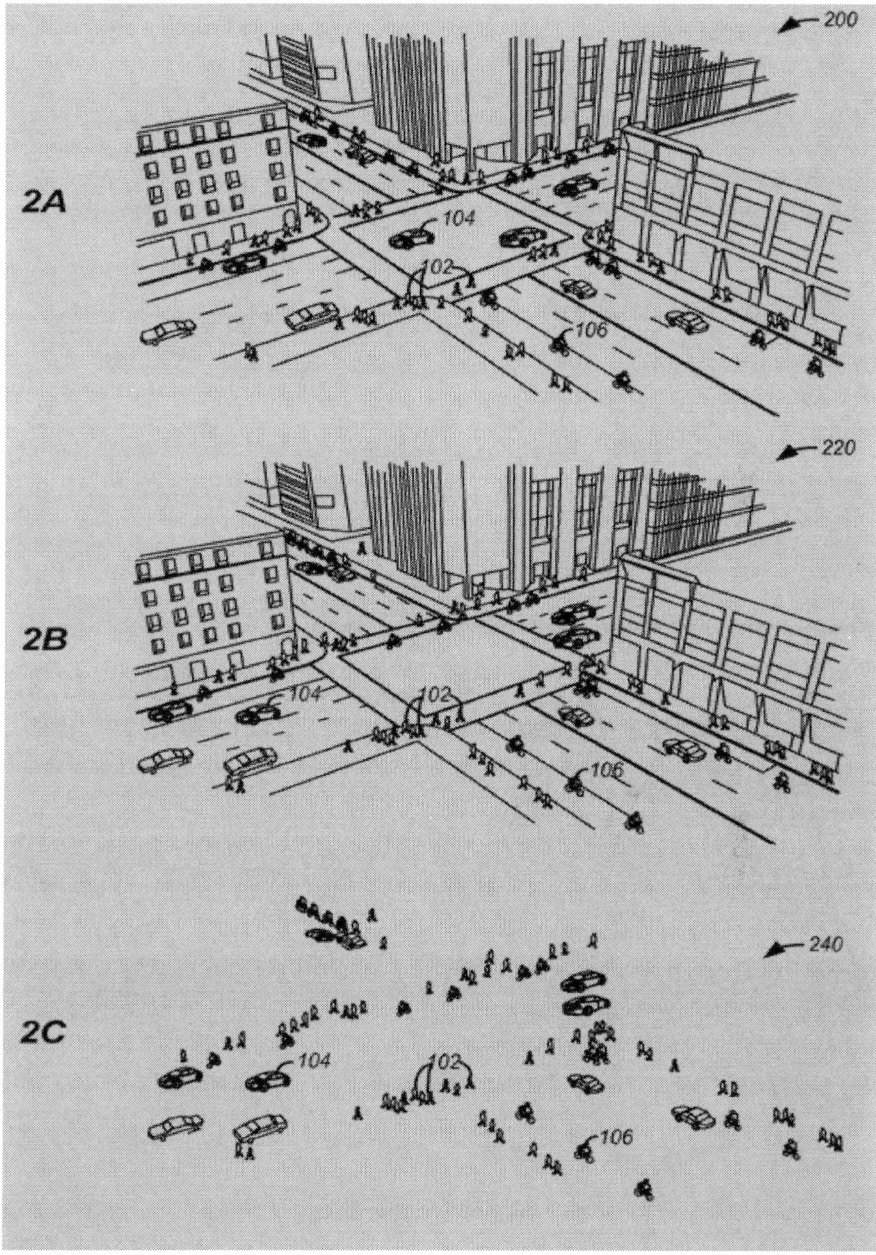

Figs. 2A, 2B, and 2C: illustrate an example process for locating foreground objects in image data that can be utilized in accordance with various embodiments.

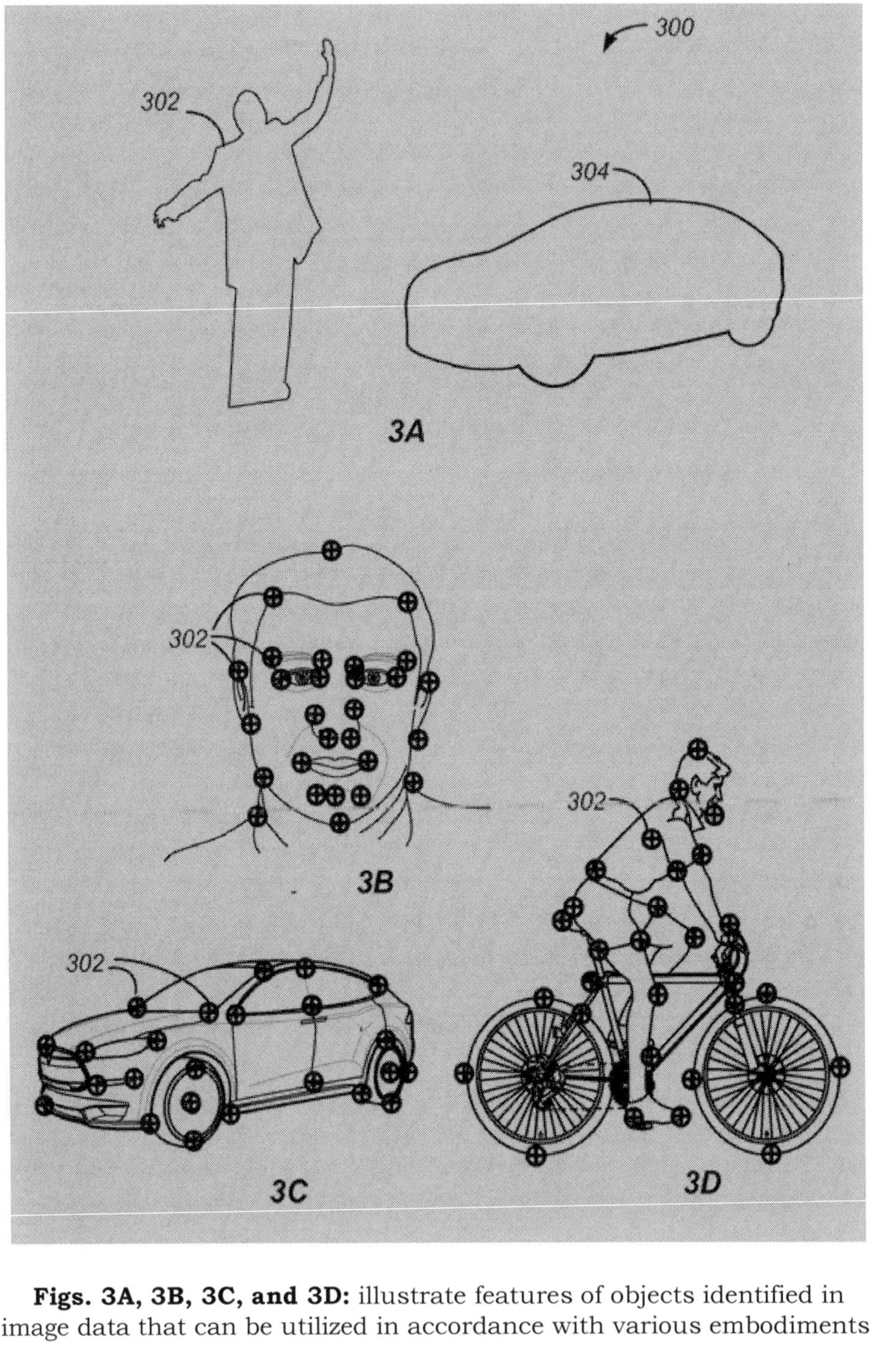

Figs. 3A, 3B, 3C, and 3D: illustrate features of objects identified in image data that can be utilized in accordance with various embodiments.

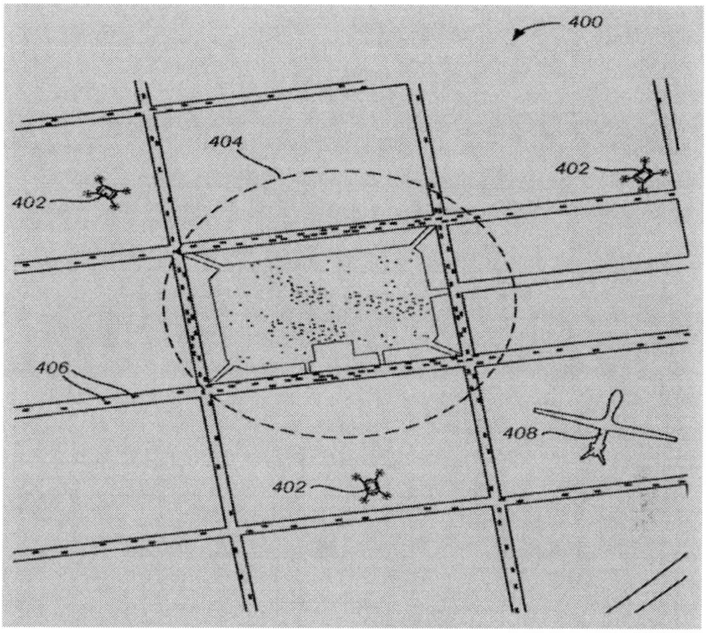

Figs 4A: is a illustrate an example approach for determining positions of interest that can be utilized in accordance with various embodiments.

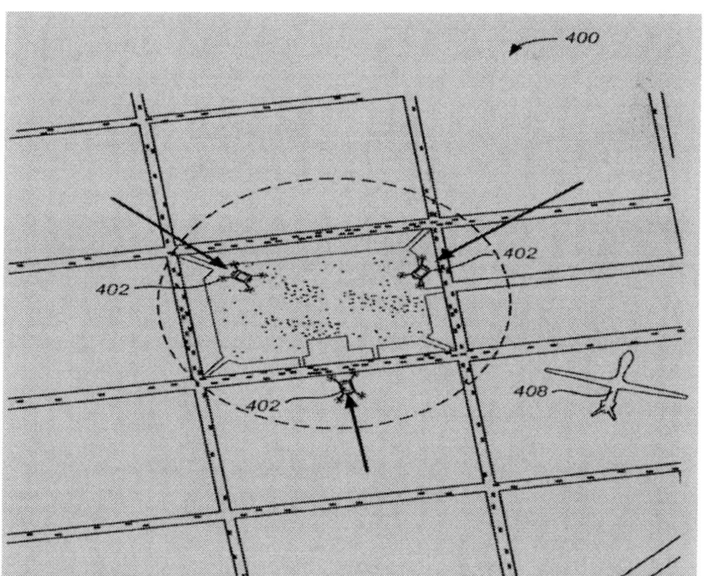

Figs. 4B: is a illustrate an example approach for determining positions of interest that can be utilized in accordance with various embodiments.

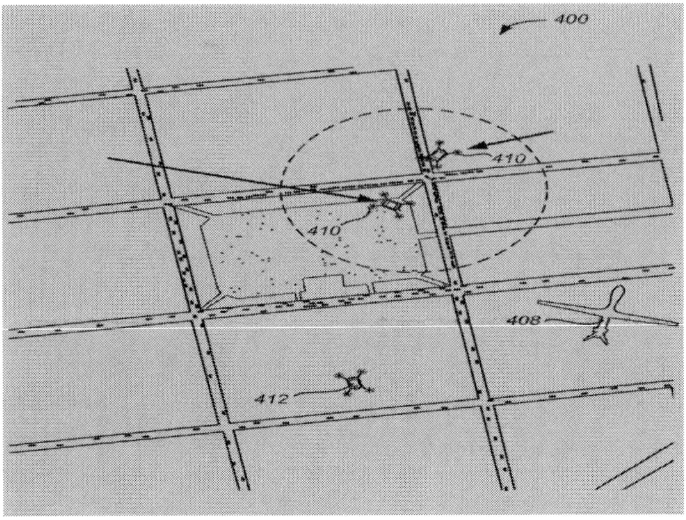

FIGS. 4C is a illustrate an example approach for determining positions of interest that can be utilized in accordance with various embodiments.

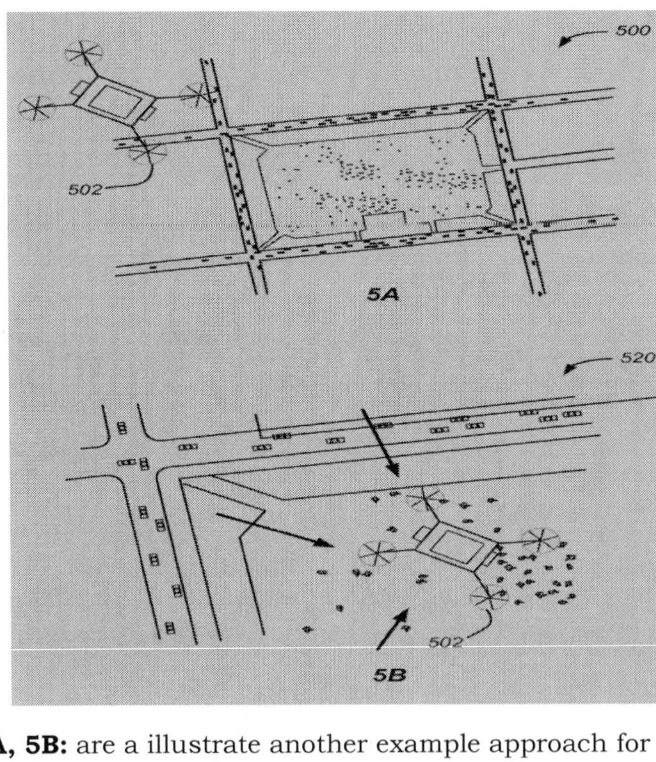

Figs. 5A, 5B: are a illustrate another example approach for obtaining additional data for specific objects or positions of interest that can be utilized in accordance with various embodiments

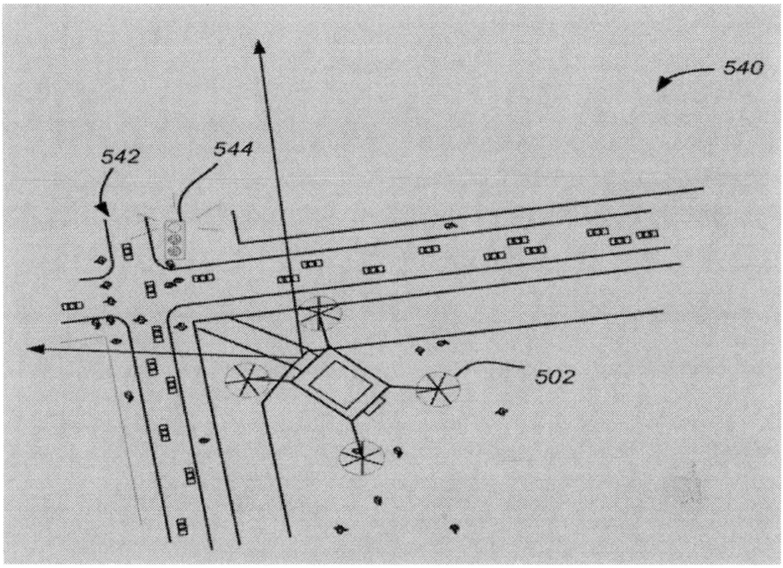

Figs. 5C: is a illustrate another example approach for obtaining additional data for specific objects or positions of interest that can be utilized in accordance with various embodiments

Fig. 6: illustrates an example situation wherein image data is captured from multiple angles to provide enhanced image data that can be utilized in accordance with various embodiments.

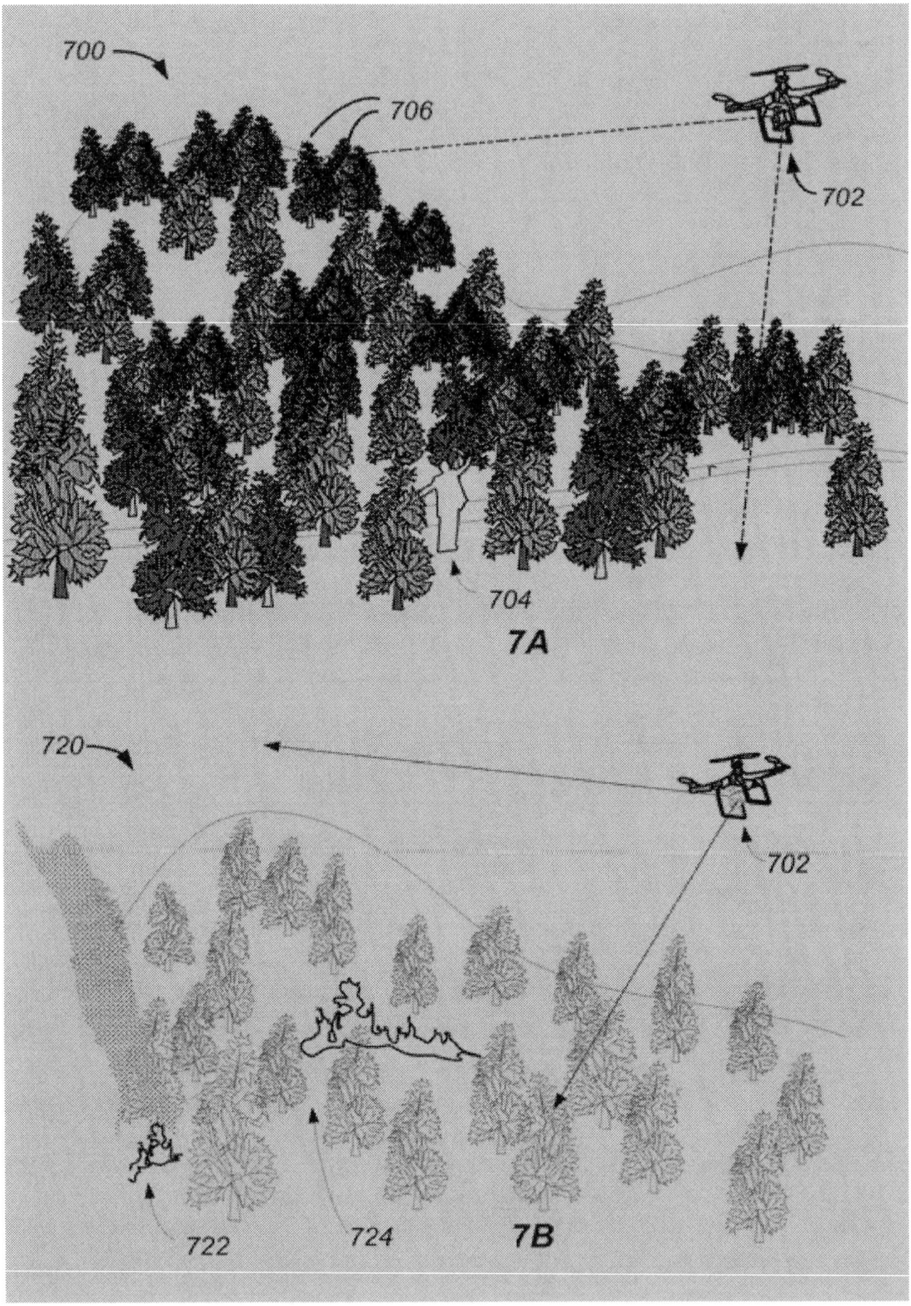

Figs. 7A and 7B: illustrate example situations wherein objects or situations can be detected in accordance with various embodiments.

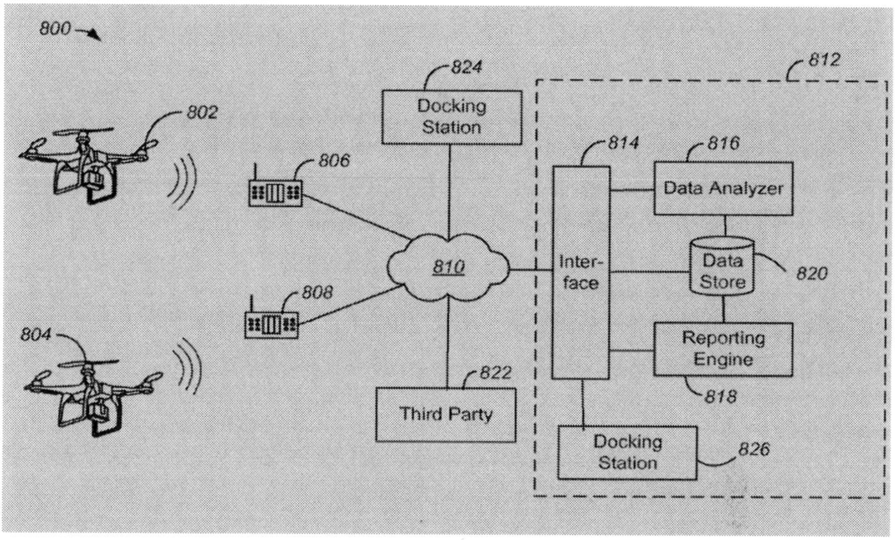

Figs. 8A and 8B: illustrate systems and devices that can be utilized to implement aspects of the various embodiments.

DESCRIPTION

Frameworks and strategies as per different epitomes of the current divulgence might conquer at least one of the previously mentioned and different lacks experienced in ordinary ways to deal with finding and recognizing actual items. Specifically, different exemplifications give instruments to finding objects of interest, like individuals, vehicles, items, logos, fires, and other perceptible articles. Different encapsulations empower these things to be distinguished, checked, followed, observed, as well as in any case represented using caught picture information. The picture information can be caught utilizing at least one automated aeronautical vehicles (UAVs), or "robots," among other such gadgets and frameworks. Different capacities and benefits are portrayed and proposed underneath as might be furnished as per the different exemplifications.

There can be numerous circumstances where it could be alluring to find objects of interest, for example, to decide the quantity of items in a given area whenever, just as to decide examples of movement, conduct, and other such data. For instance, FIG. 1 represents a climate 100 that relates to a metropolitan region, for example, may incorporate streets, crossing points, walkways, structures, and other such components. There might be many articles that move inside this climate, as may incorporate walkers 102, car vehicles 104, bikes 106, and so forth. A portion of these items might emanate sound as well as radio signs 108 or information, like discourse from an individual, a perceivable cell signal, a remote information signal (i.e., Wi-Fi or Bluetooth®) from a vehicle or cell phone, motor commotion, and so forth

A portion of these articles might hold any importance with different people or substances for assortment of reasons. For instance, a traffic office may be keen on the measure of traffic at some random time, including synopsis information like bearings of traffic stream, blockage focuses, just as varieties in rush hour gridlock over the long run. A traffic organization may likewise be keen on brief deviations from these examples, for example, may result from mishaps, occasions, or other such events. An office additionally may be keen on a breakdown of the kinds of vehicles, for example, the level of vehicles that relate to cabs or business vehicles versus buyer vehicles.

Additionally, an administrative or city office or office may be keen on acquiring such data for different kinds of traffic, for example, passerby or bike traffic. This can incorporate deciding bearings of movement, clog focuses, standards of conduct, thickness, and other such data. This data can be valuable in deciding, for instance, places for new crosswalks or bicycle paths, or changes expected to intersection lights to work on day by day development of the people on foot, bicycles, Segways, skateboards, and other such objects of interest. This data can likewise assist with recognizing areas of interest, for example, by deciding when individuals are moving towards a specific area in more noteworthy numbers than expected, or fleeing from a particular area or district, among other such deviations from standard or noticed conduct.

Notwithstanding development, traffic, and occurrence information that might bear some significance with individuals like urban areas or government organizations, such data can be helpful to different kinds of elements also. For instance, a sponsor should know the volume of traffic (auto, person on foot, or something else) that passes by a given area over the long run, to such an extent that the promoter can figure out where to put publicizing and how much that promoting is worth. Likewise, the varieties in that rush hour gridlock over the long haul can empower more designated promoting and valuing varieties for the duration of the day dependent on varieties in the measure of traffic.

An individual, for example, a retailer or restaurateur can likewise utilize such data to decide if, and where, to open a store or eatery in that climate dependent on, for instance, the measure of pedestrian activity passing a specific area, just as the occasions at which that traffic cruises by. On the off chance that the robot is conveying publicizing, for example, by towing a pennant, the robot can utilize the information to decide a flight way whereby a most extreme number of people can include the flag inside their field of view not set in stone timeframe. A methodology that can be utilized to decide this and different sorts of data in such a climate will be examined concerning FIGS. 2A through 2C. In this model methodology, picture information can be caught that incorporates at least one perspectives on a climate 200 of premium, like a metropolitan community or transportation locale. As utilized in this, picture information can allude to any of different sorts of information, as may incorporate at least one actually pictures, video information, video edges or key edges, all-encompassing pictures, and stereoscopic picture information, among other such choices. Further, while conditions, for example, urban areas and metropolitan regions are given as essential models, it ought to be perceived that provincial or rural regions, among different sorts of locales, can be used too inside the extent of the different epitomes.

In FIG. 2A, similarly as with FIG. 1, the model climate 200 incorporates countless fixed articles, as may identify with walkways, streets, structures, stoplights, traffic signs, transport stops, and such. These items that for the most part don't move after some time can be viewed as foundation objects, to such an extent that these articles can be perceived from objects of interest in the forefront that move regarding these foundation objects.

Since the objects of interest (i.e., people, canines, vehicles, bikes) will by and large move throughout some stretch of time, these items can be recognized put together basically to a limited extent with respect to their development in certain exemplifications. While approaches, for example, PC vision can be utilized to investigate a whole scene of information, for example, a picture of the whole climate 200 delineated in FIG. 2A, such methodologies can be somewhat processor and memory concentrated. On the off chance that the picture is caught by a gadget like an automated aeronautical vehicle (a UAV, alluded to hereinafter as a "drone" for straightforwardness, yet such use ought not restrict the extent of the different epitomes to explicit kinds of automated airplane or vehicles except if in any case expressed), that robot may have somewhat restricted handling and additionally memory limit, to such an extent that running a PC vision calculation on full goal, full shading video information may not be conceivable, or will be unable to be performed with adequate recurrence for to some degree a few applications.

While the video or picture information can be communicated to a far off framework or administration with adequate handling, memory, and other asset limit with respect to performing PC vision and item acknowledgment, for instance, the data transmission required may surpass the accessible association limit, or an adequate association probably won't have the option to be dependably and reliably got, among other expected issues. Further, if the broke down information is to be utilized to assist with directing the robot to explicit areas, then, at that point the inertness associated with sending and getting the information might be too incredible to possibly be valuable in different circumstances. Along these lines, it tends to be attractive to limit the measure of preparing that will be performed on the actual robot.

One approach to limit the measure of preparing to be performed on the robot (or a base station related with the robot, activities spread across a gathering of robots, and so on) is to eliminate, take away out, or in any case have the option to overlook bits of the picture information that compare to foundation items or districts. As referenced, the foundation for the most part will stay fixed in the caught picture information, essentially where the picture information is caught according to a solitary perspective.

For example, the condition of the climate 220 showed in FIG. 2B shows the areas of the objects of revenue sooner or later t after the condition of the climate 200 represented in FIG. 2A. As can be seen, most of the walkers 102, vehicles 104, and bikes have changed positions as well as directions between figures. It ought to be brought up that reference numbers might be continued between figures for motivations behind straightforwardness of clarification, however such use ought not be deciphered as a restriction on the extent of the different epitomes except if in any case explicitly expressed.

On the off chance that picture information is caught for the two conditions of the climate 200, 220, and extra picture information might be accessible for correlation, the segments of the picture information that compare to foundation still up in the air, and those can be taken out (or disregarded) from the picture information to leave staying those parts of the picture information that relate to the “forefront” objects. In the event that a bunch of pictures or video is accessible for the climate, and the climate is seen not really settled perspective, then, at that point the normal pixel esteems for every pixel area of a picture will have a normal worth that relates to the shading, force, or other such factors of the foundation at that area. Albeit a couple of pictures like those for FIGS. 2A and 2B can be utilized, a bigger examining of pictures can accommodate a more precise normal incentive for every pixel position.

On the off chance that the picture information for FIG. 2B has the normal qualities for the foundation deducted from the qualities for each relating pixel area, the leftover parts 240 of the picture information will compare to those bits of the picture information that don't compare to foundation districts or articles, as the deduction didn't bring about pixel upsides of around nothing.

It ought to be perceived that lighting, shadow, movement, force, and different elements can influence the real qualities after deduction, to such an extent that there can be an edge or reach set around a zero-pixel esteem after deduction to such an extent that pixels with pixel esteems inside a limit worth of zero will be viewed as foundation in some measure a few exemplifications. The excess parts 240 will relate to "masses" of picture information, or adjoining pixel districts, that each compare not really set in stone forefront object, or a likely object of interest. Along these lines, the PC vision, object acknowledgment, or other such cycle need just consider the areas of the forefront protests as delineated in FIG. 2C, instead of the full picture information of FIG. 2B. A benefit to taking a gander at the picture information over the long haul and playing out a normal is that the cycle will in general limit the impact of transitory impediments, lighting varieties, commotion, and other such factors that can briefly influence the pixel esteems for different areas.

Different methodologies can be utilized for reasons for foundation deduction, or frontal area identification, also inside the extent of the different exemplifications. For instance, some measure of picture pre-preparing can be performed for motivations behind working on the nature of the picture, as may incorporate sifting through commotion, changing brilliance or difference, and so forth In situations where the camera may be moving, some measure of position or movement remuneration might be proceeded also. Foundation deduction moves toward that can be used with different exemplifications incorporate mean separating, outline differencing, Gaussian normal handling, foundation combination displaying, combination of Gaussians (MoG) deduction, and such. Libraries, for example, the OPEN CV library can likewise be used to exploit the traditional foundation and closer view division calculation.

When the frontal area divides or "masses" of picture still up in the air, those bits can be handled utilizing a PC vision calculation for object acknowledgment or other such cycle. Article acknowledgment normally utilizes at least one classifiers that have been prepared to perceive explicit kinds of classifications of items, like individuals, vehicles, bikes, and so forth. Calculations utilized for such purposes can incorporate convolutional or other profound neural organizations (DNNs), as may use at least one element extraction libraries for distinguishing kinds of element points of different items.

In certain encapsulations, a histogram or situated slopes (HOG)-based methodology utilizes include descriptors for object discovery, for example, by including events of inclination direction in restricted bits of the picture information. Different methodologies that can be utilized exploit elements, for example, edge direction histograms, shape settings, and scale-invariant component change descriptors, albeit these methodologies may not give a similar degree of exactness for certain informational collections.

In certain encapsulations, an endeavor to group protests that doesn't need accuracy can depend on the overall states of the masses or frontal area locales. For instance, in the model forefront locale 300 of FIG. 3A, there are two masses that compare to various kinds of items. The main mass 302 has a framework that a classifier may show relates to a human with 85% sureness. Certain classifiers may give numerous certainty or conviction esteems, to such an extent that the scores gave may demonstrate a 85% probability that the mass relates to a human and a 5% probability that the mass compares to an auto, in view of the correspondence of the shape to the scope of potential shapes for each kind of article, which in certain encapsulations can incorporate various postures or points, among other such alternatives. Additionally, the second mass 304 has a shape that a prepared classifier could demonstrate has a high probability of relating to a vehicle. For circumstances where the items are noticeable over the long run, with the end goal that extra perspectives as well as picture information can be acquired, the picture information for different bits of each mass can be amassed, arrived at the midpoint of, or in any case handled to endeavor to further develop accuracy and certainty. As referenced somewhere else thus, the capacity to get sees from at least two unique cameras can assist with working on the certainty of the article acknowledgment measures.

Where more exact IDs are wanted, the PC vision measure utilized can endeavor to find explicit component focuses as talked about above. As specific illustrations, FIGS. 3B, 3C, and 3D delineate parts of a human, a vehicle, and a bike, individually. As referenced, various classifiers can be utilized that are prepared on various informational indexes or potentially use various libraries, where explicit classifiers can be used to endeavor to distinguish or perceive explicit sorts of items. For instance, a human classifier may be utilized with an element extraction calculation to recognize explicit component focuses 302 of a closer view item, and afterward break down the spatial relations of those element focuses to decide with no less than a base degree of certainty that the frontal area object compares to a human.

The element focuses found can relate to any provisions that are distinguished during preparing to be illustrative of a human, for example, facial elements and different elements illustrative of a human in different postures. Comparative classifiers can be utilized to decide the element focuses 302 of other forefront object to recognize those items as vehicles, bikes, or different objects of interest. On the off chance that an article isn't related to no less than a base degree of certainty, that item can be eliminated from thought, or a robot or other gadget can endeavor to acquire extra information to endeavor to decide the sort of article with higher certainty.

In certain encapsulations the picture information can be put something aside for resulting examination by a PC framework or administration with adequate preparing, memory, and other asset ability to play out a more powerful investigation. Along these lines, the less asset escalated interaction can be utilized on the robot for a fast, generally exact assurance, for example, might be valuable for deciding a flight way for the robot, while a more strong preparing can be performed by cloud workers or other such registering assets to decide a more precise check or investigation of objects of revenue in certain encapsulations.

In the wake of preparing utilizing a PC vision calculation with the fitting classifiers, libraries, or descriptors, for instance, an outcome can be acquired that is an ID of every likely object of interest with related certainty value(s). At least one certainty limits or measures can be utilized to figure out which objects to choose as the showed type. The setting of the edge worth can be a harmony between the craving for accuracy of ID and the capacity to incorporate items that give off an impression of being, yet may not be, objects of a given sort. For instance, there may be 1,000 individuals in a scene.

Setting a certainty limit excessively high, for example, at close to 100%, might bring about a check of around 100 individuals, however there will be an extremely high certainty that each article distinguished as an individual is really an individual. Setting a limit excessively low, for example, at half, may bring about such a large number of bogus up-sides being tallied, which may bring about a check of 1,500 individuals, 33% of which don't really relate to individuals. By and large, the quantity of bogus up-sides is adjusted by the quantity of people missed, with the end goal that the general tally is roughly right all things considered.

For some applications this can be a limit between about 60% and about 85%, despite the fact that as examined the reaches can change by application or circumstance. A circumstance where a robot is endeavoring to recognize the quantity of individuals in a group from 400 feet noticeable all around a use a limit that is considerably lower than for a camera at ten feet that is distinguishing the quantity of individuals in a room, in light of components, for example, goal and impacts of slight camera turns on the caught information, and so forth. The capacity to perceive particular sorts of objects of interest, like people on foot, bikes, and vehicles, empowers different kinds of information not really settled that can be helpful for an assortment of purposes. As referenced, the capacity to check the quantity of vehicles halted at a convergence or individuals in a crosswalk can assist with deciding the traffic in a specific region, and changes in that tally can be observed after some time to endeavor to decide thickness or volume as a factor of time. Following these articles after some time can assist with deciding perspectives, for example, traffic stream and marks of clog. Deciding inconsistencies in thickness, conduct, or examples can assist with distinguishing circumstances like mishaps or other startling episodes.

The capacity to get the picture information and give information in regards to perceived items could be presented as an independent framework that can be worked by offices or substances, for example, traffic offices and other legislative offices. The information likewise can be given as a feature of a help, whereby an association gathers and breaks down the picture information, and gives the information as a component of a one-time project, progressing checking venture, or other such bundle. The client of the help can determine the kind of information wanted, just as the recurrence of the information or length of checking, and can be charged in like manner. In certain epitomes the information may be distributed as a feature of a membership administration, whereby a portable application supplier or other such substance can get a membership to distribute or get the information for purposes, for example, route and course assurance. Such information likewise can be utilized to assist with recognizing mishaps, development, clog, and other such events.

As referenced, a large number of the models thus use picture information caught by at least one cameras with a perspective on a space of interest. In certain encapsulations these cameras can be given utilizing at least one robots, which can each incorporate at least one cameras or different sensors for catching the picture information. These can incorporate, for instance, computerized still cameras, advanced camcorders, infrared finders, stereoscopic cameras, warm sensors, movement sensors, vicinity sensors, and other such sensors or parts.

The picture information can incorporate at least one pictures, or video, demonstrating pixel esteems for pixel areas of the camera sensor, for instance, where the pixel esteems can address information like the power or shade of surrounding, infrared IR, or bright (UV) radiation recognized by the sensor. A robot may likewise incorporate non-visual based sensors, like radio or sound collectors, for recognizing energy exuding from different objects of revenue. These energy sources can incorporate, for instance, mobile phone signals, voices, vehicle commotions, and so forth. This can incorporate searching for particular signs or an absolute number of signs, just as the data transmission, blockage, or throughput of signs, among other such alternatives. Sound and other mark information can assist with deciding viewpoints like kind of vehicle, locales of movement, and so forth, just as giving one more contribution to tallying or following purposes. The general sound level and course of the sound can likewise give an extra contribution to possible areas of interest.

A robot might incorporate different cameras or sensors also, as might be utilized for purposes, for example, crash evasion, way assurance, and flight control, among other such errands. In some measure a portion of the information from these sensors might be utilized to endeavor to expand exactness of the article assurance measure. As referenced somewhere else thus, the sensors can likewise incorporate movement and direction sensors that can assist with representing evident development in the picture information that is an aftereffect of development of the camera or robot used to catch the picture information.

While catching the picture information as a component of an observing cycle, for instance, at least one robots can be arranged to float over a space of interest, or follow a standard flight way finished, or around, the space of interest. Contingent on the sort of region being noticed, just as different factors like the size or weight of the robot, there may be a particular stature or distance that the robot needs to keep up with inside the space, like a tallness somewhere in the range of 100 and 200 feet, or 200 and 400 feet, albeit different statures are being proposed and may fluctuate by area and ward, and so forth It might then be significant for certain applications that the robots have the option to move to an ideal area, or if nothing else a fitting area, from which to catch picture information for different articles or events.

In like manner, approaches as per different exemplifications use in some measure some measure of item acknowledgment and investigation on the actual robots to decide changes in position and flight way that can help with acquiring suitable picture information. This can incorporate cycles, for example, forefront location and moderately basic item acknowledgment to recognize data like thickness of articles, object development and stream, and so forth This information doesn't need to be inconceivably precise, yet adequate to empower the robot to decide an event of a kind that the robot ought to research.

For instance, FIG. 4A outlines a model circumstance 400 wherein a gathering comprising of three rotating drones 402 and one fixed-wing drone 408 is arranged to screen objects of interest in an assigned district, here a nine square space of a particular city. Each robot can fly at a generally undeniable level for reasons for checking, to perceive objects 406 like vehicles and people on foot, and decide when there is a huge change in thickness, volume, stream, or other such aspect(s). In this model a fixed-wing drone 408 flies at a most elevated height to acquire an outline of a generally huge region, while every revolving drone 402 can fly at a lower elevation to get information over a particular segment of that space.

The robots 402, 408 can each do their own assessment, can share information, or can in any case impart and exploit the perspectives or conclusions of different robots. In this model, there is an expansion in number and thickness of individuals in a space 404 related with the fundamental court. Likewise, any or the entirety of the robots seeing this change can, as a component of a criticism circle or other such element or cycle, decide if the change warrants further examination as per different limit, rules, or arrangements set up for the robot control frameworks. In this model, one of the robots discovers that there is a change that meets a further examination standard, and communicates something specific, demand, or other arrangement of data to different robots, either explicitly recognized or not really settled locale, in regards to the event. In certain epitomes this can cause at least one of the robots to move to an area, or along a flight way, whereby the drone(s) will actually want to get picture information for sees, or further developed perspectives, of the space.

In certain epitomes one of the robots, for example, the fixed-wing drone 408 in this model, might be a "ace" drone that can make different robots move to acquire explicit perspectives or follow explicit ways. In certain encapsulations any robot can turn into the expert for a given circumstance, in view of elements like area, height, arrangement, mobility, and such. The expert robot might keep a specific elevation, or might be a bigger robot that will most likely be unable to move as near objects of interest, among other such alternatives.

On the off chance that individuals or items in a space are thickly stuffed in to such an extent that development is insignificant, it very well might be hard to utilize lightweight foundation deduction techniques on a robot. In these circumstances, at least one of the robots might communicate picture information to a distant framework for investigation and item distinguishing proof, and so on The distant framework can likewise send back data about the distinguished foundation areas all together for at least one of the robots to refresh their classifiers or preparing information, or in any case have the option to recognize foundation all the more precisely for motivations behind flight way assurance and different angles examined thus.

In the model circumstance 400 of FIG. 4B, the robots 402 arrange (either by and large or under bearing of the "ace" drone 408) to move to positions whereby the robots 402 can each acquire picture information for perspectives on the court where the expanded still up in the air to happen. In this model, each robot is appointed to a particular piece of the space, to such an extent that various perspectives on the court can be acquired without pointless cross-over or making the robots fly also near one another.

While the robots can be outfitted with crash evasion frameworks and other such parts, the robots can likewise be arranged to fly dependent on fields of perspective on their particular cameras, to such an extent that a most extreme measure of extraordinary information can be acquired from the robots that are available. Different methodologies can be utilized too, for example, to make drones catch covering or repetitive information for motivations behind further developing exactness and accuracy, and so forth As referenced somewhere else in this, the various perspectives can likewise be used to get three-dimensional information as well as different perspectives on objects of interest.

FIG. 4C shows a comparable model circumstance in a similar area wherein gridlock is distinguished in a particular district. The recognition can be the consequence of a thickness of vehicles not really settled edge, the traffic design falling external a typical reach, a measure of commotion coordinating with a blockage design, or potentially other such factor(s). In this model, the expert robot 408, which may have more preparing and memory limit than the rotational robots, can decide to send a subset of the robots 410 to research the occurrence or inconsistency. While all robots in the space can be coordinated to acquire extra data in certain encapsulations, in this model the expert robot 408 decides the quantity of robots to move to that space dependent on components like the size of the space of the episode, the kind of occurrence, the general area of each robot, the certainty that the episode is a significant occurrence, and so forth The robots 410 would then be able to move not really settled area and catch picture/sensor information for the area, which can be broke down on the robots 410 themselves, conveyed to the expert robot 408, as well as imparted to a distant framework for investigation. As blockage expands, diminishes, or changes in area, robots can be redistributed as suitable.

The robots 410 can utilize data about the course of the clog, and development of vehicles in different ways, to decide the probable area of the reason for the blockage to endeavor to get extra perspectives and data for, and around, that area. One of the other rotating drones 412 that isn't coordinated to examine the occurrence can proceed with its present way or can be coordinated to float in a particular area to get outline data, among other such choices.

Indeed, even in circumstances without numerous robots, a solitary robot can use object conclusions to change the flight way or decide regions for additional examination. For instance, in the circumstance 500 of FIG. 5A a solitary robot 502 may be adhering to its standard flight way over a city, then, at that point verify that there is uncommon movement in a specific region or area. In certain encapsulations the robot might be arranged to consequently move towards that space to get extra data. In certain encapsulations the robot may initially break down the information to decide if the data acquired for the space meets a base certainty limit, to such an extent that the articles are being related to an adequate degree of exactness.

Provided that this is true, the robot can proceed with its flight way or in any case keep up with its present height. On the off chance that the data got doesn't meet a limit or models, or then again in case there is an alternate edge for surprising movement, the robot can investigate the circumstance and decide a fitting area or way to continue to acquire extra data to expand the certainty esteems. This can incorporate, for instance, meaning the suitable region (or as close as could be expected, regarding distance guidelines and restricted air spaces, for instance) and dropping to a height, for example, almost a base reasonable elevation, whereby the robot can catch a bigger, higher goal perspective on the articles as represented in the model circumstance 520 of FIG. 5B. As referenced, in certain encapsulations the robot might be restricted in its vicinity to people, for instance, with the end goal that a fitting area or flight way ought not set in stone to empower the camera to get as near the items as could really be expected, or possibly down to earth, while noticing every single relating guideline.

In certain encapsulations a robot can be arranged to screen a specific area or sort of area. For instance, in the model circumstance 540 of FIG. 5C a robot 502 is told to screen at least one convergences throughout some undefined time frame. This observing can incorporate deciding the presence and areas of items like people on foot, bikes, and vehicles, and following their development. The checking can likewise include deciding variables, for example, the condition of a traffic light or cross-walk signal. The robot can examine the caught data to decide when walkers or bikes are crossing with or against the light, just as how quick vehicles are traveling through the convergence and the number of vehicles go through the convergence against the light.

This data can decide possibly hazardous convergences, and the manners by which those crossing points are risky. For instance, the robot information may demonstrate that vehicles a first way at a first convergence will in general speed through the crossing point, and at a second convergence walkers and bikes may cross against the light more than ordinary. What's more, these examples may differ by season of day and day of the week, in view of when individuals are driving, and so on This data would then be able to be examined, for example, by a backend framework, to recognize potential risks to such an extent that instruments can be set up to decrease those risks and forestall mishaps or other such undesired events. In certain encapsulations this can be given as a feature of a checking administration, whereby a city or local area can demand the observing be done throughout some stretch of time for a specific region or set of areas, and the subsequent information or examination of the information can be given at fitting occasions.

It ought to be noticed that straightforwardly overhead perspectives may not be ideal in certain circumstances. For instance, it could be hard to distinguish an individual wearing a cap or holding an umbrella when flying straight over that individual. A view like a side or viewpoint see, which can give to some extent some element focuses for the client's face or posture, perhaps liked in to some degree a few exemplifications. As needs be, the flight control programming may decide the areas of the objects of interest, then, at that point decide the proper flight intend to use to acquire the ideal perspectives on those objects of interest. Further as referenced, in certain exemplifications it very well might be attractive to catch perspectives on objects from various points to further develop certainty levels. For instance, a perspective on an individual from the front, where the face is noticeable, may give the most elevated level of certainty for a human article where a view from the side may be the ideal area for a bike, to such an extent that it very well might be attractive to acquire different perspectives, or if nothing else move around an object of interest until a base certainty level is gotten.

Numerous perspectives can be acquired utilizing a solitary robot with a suitable flight way. While this can assist with recognizing objects, the time expected to move to the different perspectives can empower the items to move, turn, or leave the region to such an extent that it may not be feasible to acquire the different perspectives. Appropriately, for a solitary robot the framework can be arranged to use as much information and however many perspectives on an item as might be accessible. In the event that at least two robots 606, 608 are accessible for a space, as shown in the model climate 600 of FIG. 6, those robots can be arranged to simultaneously catch picture information addressing various perspectives on at least one objects of interest. The information can be time synchronized with the end goal that the relating perspectives can be associated, empowering more precise assurance of articles and conceivably the age of three-dimensional models or portrayals of the items.

Further, the capacity to acquire two unique perspectives from definite perspectives empowers the catching of stereoscopic picture information, whereby the divergence gives data about distance and scale. For instance, individuals 602, 604 in FIG. 6 are both contained inside the fields of view 610, 612 of the cameras of the separate robots 606, 608. In the event that, utilizing sensors, for example, position sensors, geographic sensors, altimeters, closeness sensors, as well as other such sensors, the robots 606, 608 can decide their relative positions, the information caught by every camera can be corresponded to furnish dissimilarity data that fluctuates with distance. By knowing the optical properties of the camera congregations and relative positions, the divergence information can be utilized to acquire distance and measurement information for individuals 602, 604 and different articles in the scene. The measurement information can assist with expanding the trust in the objects of interest. For instance, if the item being recognized is a doll or activity figure that is just three inches tall, then, at that point despite the fact that the element acknowledgment cycle may distinguish the activity figure as a human with a base degree of certainty, the measurements can demonstrate that the article is really not a grown-up human.

Essentially, if the item is a twelve-foot-tall sculpture, the measurement information can be utilized to preclude the chance, basically with an extremely high probability, that the article is a grown-up human. Different elements can be utilized to further develop trust in different epitomes also, for example, shading investigation and so forth. For instance, on the off chance that a sculpture object is monochrome bronze, the sculpture can be precluded as a human with high probability regardless of whether the component focuses match a human item as indicated by the proper classifier. Different sensors that are fit for distinguishing heat marks or different information characteristic of a human item can be utilized to build certainty too inside the extent of different exemplifications.

The capacity to rapidly perceive objects from a robot, just as the capacity to use various kinds of information to make those conclusions, can offer critical benefit in different circumstances also. For instance, in the circumstance 700 of FIG. 7A a robot 702 is flying over a woods as a feature of an inquiry and salvage activity. For this situation, the thickness of the trees can make it hard for the robot to get an adequate perspective on an individual to make the assurance with a serious level of certainty dependent on picture information from a flight elevation alone. Utilizing approaches talked about and recommended thus, the robot could fly over the space catching picture information to do fast closer view detachment and picture investigation to endeavor to decide any items that match a human with even an exceptionally low certainty level, for example, might be sufficient to keep away from commotion and inordinate bogus up-sides however adequate to distinguish an expected human when just a part of that human is noticeable. In security tasks, the robot may not be searching for a specific individual yet any individual in a space at a specific time, whereby the robot can play out an activity, for example, to endeavor to perceive the individual, send picture information to security staff, sound an alert, or play out another such activity.

Notwithstanding picture information, heat signature information can be caught and investigated simultaneously to endeavor to work on the trust in human identification judgments. A generally low degree of certainty, for example, essentially 20% certainty, might be adequate to cause the robot 702 to change a flight way to have the option to draw nearer to the expected object of revenue 704 and catch a bigger or potentially higher goal perspective on the item. The capacity to perform fast item acknowledgment on the robot and consolidate this with heat signature or other such information can empower the robot to filter an enormous region and move into any districts that contain an expected object of interest. The robot can proceed toward a path towards the article 704 until the certainty level dips under a base sum or the item is recognized as an alternate sort of item. In different exemplifications, the robot can proceed towards the article 704 essentially until adequate time as the item is related to an adequate degree of certainty that the item has been perceived personally. In certain epitomes a facial acknowledgment cycle or other such interaction can be utilized to endeavor to really recognize the found individual as the individual of interest.

On the off chance that the individual is recognized as the individual of interest, a suitable move can be made, for example, to send a solicitation for salvage or possibly send a notice that the individual has been found. The directions of the individual as well as robot likewise can be sent in certain epitomes. High goal picture information can be moved to a far off framework to assist with the assurance also, where that framework might have seriously preparing power and further developed facial acknowledgment calculations, or other such assets. In certain encapsulations where the robot is of adequate strength and power, and where the individual can be moved, the robot can likewise conceivably be caused to lift the individual out of the space and transport them to an assigned area.

The capacity to use heat marks and train classifiers (or HOG descriptors, and so forth) for specific kinds of articles can empower different sorts of conclusions and uses too. For instance, in the circumstance 720 of FIG. 7B there are a couple of secluded flames 722, 724 consuming in a distant wood. The capacity to have drones flying over the woodland and rapidly identify little flames before long they light can assist the fires with being battled before those flames become unmanageable. On the off chance that conceivable, the robot can likewise possibly dump dousing material to extinguish the fire while such a methodology might in any case be powerful. Likewise, with different epitomes examined over, the robot can likewise discuss remotely with different robots to make those robots move to the space to assist with dropping smothering material too. In the event that such capacity is absent, the robots can essentially catch extra picture data (as may incorporate visual and warmth signature information) to assist with giving data about the present status of the fire to assist with directing firemen to the proper areas.

Classifiers can be prepared to identify smoke, flares, heat marks, or different items that might be demonstrative of flames or other such events. In to some degree a few epitomes the robots can be coordinated to move towards the space until adequate data is accessible to decide the size of the fire, just as the presence of people close to the fire and the sort of action of those people, to decide if the fire is logical a pit fire or other contained fire that doesn't need consideration. In to some degree a few encapsulations picture information for such a fire might be communicated to a particular area or address for audit by a human just to guarantee that the fire doesn't need any sort of true activity. Robots can likewise be composed to fly nearby around the fire to recognize streams, precipices, or different elements that may normally stop the spread of the fire a specific way, to assist with directing the fire groups to areas where they can be best.

The capacity to find waterways and bluffs can likewise assist with directing the firemen along a way where they will actually want to pass without critical exertion or deviation required. A robot can investigate expected ways and convey whether ground access is conceivable or regardless of whether just air access is conceivable. The robots can likewise check the region for people, structures, or different articles that may be in danger from the fire(s).

As referenced, in different exemplifications there can be some handling performed on the robots, while a portion of the preparing can be performed utilizing a far off framework or administration. An illustration of such a climate 800 is delineated in FIG. 8. In this model, two robots 802, 804 are each ready to catch picture information to be prepared by a help offered utilizing assets of an asset supplier climate 812, for example, a cloud climate presented by an asset supplier. The robots might be of something very similar or various sorts, worked by something very similar or various substances and giving comparative or various kinds of data from comparative or various regions.

Each robot can discuss remotely with at least one base stations 806, 808, either the equivalent or diverse station worked by something very similar or various elements, which can make data be communicated across somewhere around one organization 810 to be gotten by an interface layer 814 of the asset supplier climate 812. The organization 810 can be any proper organization, like at least one wired or remote organizations as may incorporate a cell organization, the Internet, an intranet, a neighborhood (LAN), and such. The information communicated can incorporate any suitable information talked about or recommended thus, as may incorporate picture information, position information, sensor information, identifier information, object portrayal information, and so forth.

The information can be gotten to an interface, for example, an application programming interface (API) or other such interface of the interface layer 814. The interface layer can incorporate any of various parts known for getting and directing data in a climate, as may incorporate different switches, load balancers, network switches, and other such parts. The information can be put away to an information store 820 or other such archive for investigation by an information analyzer 816 or other such framework, administration, part, application, or gadget. In certain exemplifications where the information identifies with routine observing, the information might be examined at explicit occasions or stretches. In exemplifications where the information identifies with explicit occasions or events, or regardless of whether in any case showed by the information or message got from the robot, the information analyzer may stack the information into occupant memory and break down the information promptly, or quickly inside the setting of the climate.

The information analyzer can perform different insightful errands talked about and recommended thus, for example, to include objects of revenue in specific areas after some time, look for designs in object development, perform more precise article conclusions, and so forth. For routine observing and other non-basic information, the outcomes can be put away back to the information store 820 (or an alternate information store or archive) for ensuing recovery by a part, for example, an announcing motor 818, which can total the important information and produce a report or other such yield that can be given to an outsider 822 or other element who might have bought in to the assistance or in any case paid or got the rights for such a report. For basic information, for example, for a fire, mishap, or other such event, the revealing motor may be advised by the information analyzer immediately, to such an extent that the detailing motor can rapidly send a warning to the suitable element or substances, like a police division, emergency vehicle dispatch, local group of fire-fighters, and so forth

Different reports can be produced just as would be clear to one of standard ability in the workmanship considering the lessons and ideas contained in this. The information can be put away after some time for resulting examination and correlation, as might be helpful in deciding patterns or examples that can be put together essentially to a limited extent with respect to chronicled conduct information or other such data.

In certain encapsulations a robot 802 can be gotten to a docking station, which can be an interior docking station 826 of the asset supplier climate or an outside docking station 824 that might be given by a client or an outsider. In the docking station the robot can re-energize its batteries and have any suitable support or other such help performed. The robot additionally can interface with the docking station to send information to and from the information analyzer 816, information store 820, or other such area. In certain encapsulations the robot will transfer all its picture and sensor information to the asset supplier climate over a wired information association of the docking station.

The robot can likewise get information from the climate 812, like new arrangement information, prepared models, flight ways, and so forth. The robot likewise can have different errands performed while in the docking station, for example, may incorporate information evacuation, infection checking, programming refreshing, and such.

A robot, distantly steered airplane (RPA), or automated ethereal vehicle (UAV) can incorporate different cameras, sensors, and different parts that empower the robot to catch and send the pertinent information. An illustration of such a robot 850 is outlined in FIG. 8. In this model, the robot is a multi-rotor, fixed-pitch airplane, for example, a quadcopter with four separate engines and rotors 854, albeit fixed wing plans with dispatch systems and different kinds of airplane can be utilized also inside the extent of the different encapsulations.

The technique for flight and kinds of motors, flight components, and dispatch instruments are not basic to a portion of the encapsulations, as numerous exemplifications rather depend just on the robot being able to move not set in stone area as might be dictated by a proper flight control framework. While the robot can be directed by controller in certain exemplifications, in different epitomes the robot is controlled independently by a locally available PC as well as control framework.

In this model the robot has a lodging 852 made out of a lightweight material like a polymer, fiberglass, or carbon fiber material, albeit different materials can be used too that accommodate strength and sturdiness while adding as little weight as could be expected. The robot will incorporate somewhere around one camera 856 for catching picture information inside a catch scope of the robot. The camera can be any fitting camera, as may incorporate an integral metal-oxide-semiconductor (CMOS), charge coupled gadget (CCD), or other such sensor or identifier equipped for catching light energy still up in the air range, as may incorporate bits of the apparent, infrared, as well as bright range.

The camera might be important for a get together that incorporates suitable optics, focal points, centering components, shades, and other such components for picture catch by a solitary camera, set of cameras, stereoscopic camera, or other such arrangement. The camera can be designed to perform assignments, for example, self-adjusting, zoom (optical or computerized), brilliance and shading changes, and so forth.

As referenced, the robot can incorporate some measure of memory 866 for putting away somewhere around a part of the picture information, briefly on the robot. The information can be put away essentially until such time as at least one processors 860 of the robot can examine something like a part of the information and additionally somewhere around a piece of the information can be conveyed, by means of a remote specialized gadget 870, to a base station or distant framework or worker for investigation. The memory can incorporate any suitable memory, or mix thereof, as may incorporate glimmer memory or another such non-transient, non-unpredictable PC lucid capacity medium. The memory or potentially information stockpiling on the robot may likewise incorporate gadgets, for example, memory cards, USB streak drives, strong state drives, RAM, optical drives, attractive drives, and so forth.

To work the robot, the processor can work with a flight control framework 868 that can be accused of deciding the heading of trip for the robot and conveying the suitable messages to the motor(s) for the rotors 854 or other flight component to make the robot move the ideal way. The control framework can use flight way information put away in memory 866, just as unique choices made by the processor(s) 860 or guidelines got over the remote specialized gadget 870, among other such choices. The motor(s) can be any suitable engine utilized for such airplane, as may incorporate brushless engines for more modest robots to more ordinary airplane engines for bigger airplane. There can be different other drive instruments, control frameworks, adjusting parts, or different components utilized with such frameworks.

The model robot 850 additionally incorporates at least one movement or potentially direction sensors 864. These can incorporate, for instance, spinners, electronic compasses, inertial sensors, accelerometers, magnetometers, and such. These sensors can be utilized to decide movement of the gadget, which can assist with giving input to the control framework 868 just as to give data to the processor(s) 860 with regards to changes in area and direction of the robot, which can be utilized to eliminate, or represent, obvious movement in the picture information because of the robot. The camera 856 may incorporate picture adjustment parts and additionally programming, yet these lone record for little vibrations, shaking, or variety in perspective. By knowing changes in direction and position from the proper sensor(s), changes in position of articles in the picture information can be represented when those progressions are because of movement of the robot.

This can assist with deciding foundation districts, for instance, where the foundation division measure depends on foundation that doesn't fundamentally change after some time however because of movement of the robot the foundation would change in the caught picture information. The capacity to take away out the impacts of the movement, for instance, can empower the foundation picture information to be distinguished as not moving over the long haul. For instance, if the robot deciphers one way that would cause a ten pixel change a first course in the picture information, the position data in the picture information can be moved by ten pixels the other way to represent the development.

A comparable impact would be knowledgeable about revolution, where the precise change would influence the picture information not really settled rakish degree. In the event that the surmised distance to the objects of interest is referred to, for example, from an altimeter or other such part, then, at that point the picture information can be repaid as needs be. Different other movement changes can be performed on the picture information as would be evident to one of customary expertise in the craftsmanship considering the lessons and ideas contained thus.

The robot can incorporate different parts too, for example, at least one battery-powered batteries 862 or other such force sources, as may get power from sunlight based, remote, or other such sources while in flight. The robot can likewise incorporate a nearness sensor 858, ultrasonic sensor, or other sensor equipped for recognizing the closeness or area of objects to aid crash aversion. Different parts can be utilized also, for example, may incorporate autopilot frameworks, velocity sensors, propeller balancers, crash support components, and such.

OUTCOME

Our Invention "Human Detection and Motion Analysis from an Unmanned Arial Vehicle utilizing Deep Learning Techniques "is an automated ethereal vehicle (UAV) can incorporate more than one cameras for catching picture information inside a field of view that depends to some degree upon the area and direction of the UAV under profound learning procedure. The innovation additionally a piece of the picture information can be handled and investigation on the UAV to find objects of revenue like moving items, individuals, leaving things or vehicles, and utilize that data to figure out where to fly the robot to catch better picture information of those or other such articles. The imagined innovation additionally distinguished the objects of interest can be checked and the thickness, development, area, and conduct of those articles recognized and furthermore this can assist with deciding events, for example, gridlock or surprising examples of walker development, just as to find people, flames, or other such items. The creation likewise be dissected by a distant framework or administration that has extra assets to give more exact outcomes approx. 98%.

1. According to claim1# the innovation is to an automated airborne vehicle (UAV) can incorporate more than one cameras for catching picture information inside a field of view that depends to some degree upon the area and direction of the UAV under profound learning procedure.
2. According to claim1,2# the innovation is to a piece of the picture information can be prepared and examination on the UAV to find objects of interest like moving articles, individuals, leaving things or vehicles, and utilize that data to figure out where to fly the robot to catch greater picture information of those or other such items.
3. According to claim1,2,3# the innovation is to a distinguished the objects of interest can be tallied and the thickness, development, area, and conduct of those items recognized and furthermore this can assist with deciding events, for example, gridlock or strange examples of walker development, just as to find people, flames, or other such articles.
4. According to claim1,2,3,4# the innovation is to the creation likewise be examined by a distant framework or administration that has extra assets to give more exact outcomes approx. 98%.

REFERENCES

[1] Abhigna, B.S., Chakraborty, S., Kumar, V., Medha: A collaborative E-concept framework, Proceedings of the 2018 International Conference On Communication, Computing and Internet of Things, IC3IoT 2018, 2019, pp. 117–122, 8668127

[2] Abidin, S., Vadi, V.K., Rana, A., On Confidentiality, Integrity, Authenticity, and Freshness (CIAF) in WSN, Advances in Computer, Communication and Computational Sciences, 2020, pp 87-97

[3] ahay, A., Choudhury, T., Kumar, V., Hierarchical Evolution of Digital Games, Proceedings of 2019 International Conference on Computational Intelligence and Knowledge Economy, ICCIKE 2019, 2019, pp. 381–386, 9004378

[4] Area: Communication, Frequency modulation, Networks, wireless communication

[5] Area: Communication, Frequency modulation, Networks, wireless communication

[6] Area: Communication, Frequency modulation, Networks, wireless communication

[7] Area: Communication, Frequency modulation, Networks, wireless communication

[8] Arora, M., Kumar, V., Dave, M., Hybrid cloud infrastructure for load balancing and scaling with vm path optimization using ant colony optimization, Journal of Advanced Research in Dynamical and Control Systems, 2020, 12(5 Special Issue), pp. 660–674

[9] Arora, M., Kumar, V., Dave, M., Task scheduling in cloud infrastructure using optimization technique genetic algorithm, Proceedings of the World Conference on Smart Trends in Systems, Security and Sustainability, WS4 2020, 2020, pp. 788–793, 9210303

[10] Bansal, R., Gupta, A., Singh, R., Nassa, V.K., Role and impact of digital technologies in E-learning amidst COVID-19 pandemic, Proceedings - 2021 4th International Conference on Computational Intelligence and Communication Technologies, CCICT 2021, 2021, pp. 194–202

[11] Bansal, R., Obaid, A.J., Gupta, A., Singh, R., Pramanik, S., Impact of Big Data on Digital Transformation in 5G Era, Journal of Physics: Conference Seriesthis link is disabled, 2021, 1963(1), 012170

[12] Chauhan, S.S., Khare, A.K., Kinematic analysis of the ABB IRB 1520 industrial robot using roboanalyzer software, Evergreenthis link is disabled, 2020, 7(4), pp. 510–518

[13] Chauhan, S.S., Khare, A.K., Method of Obtaining Planar State of Stress Using Mohr's Circle—Some Typical Cases, Lecture Notes in Mechanical Engineering, 2022, pp. 33–38

[14] Goel, S., Tripathi, A., Choudhury, T., Kumar, V. Parkinson's Disease Detection using Soft Computing Technique, Proceedings of the 2019 8th International Conference on System Modeling and Advancement in Research Trends, SMART 2019, 2020, pp. 3–6, 9117384

[15] Govindarajan, V., Sivakumar, R., Patil, P.P., Kannan, M., Ramesh, B., Effect of Tungsten Carbide Addition on the Microstructure and Mechanical Behavior of Titanium Matrix Developed by Powder Metallurgy Route, Advances in Materials Science and Engineeringthis link is disabled, 2022, 2022, 2266951

[16] Gupta, A., Singh, R., Nassa, V.K., Sharma, P., Koti, K., Investigating Application and Challenges of Big Data Analytics with Clustering, 2021 International Conference on Advancements in Electrical, Electronics, Communication, Computing and Automation, ICAECA 2021, 2021

[17] Jain, P., Vijay, S., & Gupta, S. C. (2009). ATM Network QoS Parameters and Performance Analysis using OPNET Simulator. International Journal of Recent Trends in Engineering, 2(4), 228.

[18] Jain, P., Vijay, S., & Gupta, S. C. Performance Analysis & QoS Guarantee in ATM Networks. Global Journal of computer Science and Technology, 131-136.

[19] Jain, S., Choudhury, T., Kumar, V., Kumar, P., Detecting malware and analysing using sandbox evasion, Proceedings of the 2018 International

Conference On Communication, Computing and Internet of Things, IC3IoT 2018, 2019, pp. 111–116, 8668149

[20] Kala, S., Nandan, H., Sharma, P., Shadow and weak gravitational lensing of a rotating regular black hole in a non-minimally coupled Einstein-Yang-Mills theory in the presence of plasma, European Physical Journal Plusthis link is disabled, 2022, 137(4), 457

[21] Kandpal, H., Kumar, A., Malik, A., Morphometric analysis and prioritization of sub-watersheds in Bino watershed, Uttarakhand: A remote sensing and GIS perspective, Indian Journal of Ecology, 2018, 45(3), pp. 475–478

[22] Kannan, M., Singh, S., Prasad, R.R., Synthetic Methods to Obtain Calcia-Stabilized Zirconia Powders—A Review, Lecture Notes in Mechanical Engineeringthis link is disabled, 2021, pp. 405–416

[23] Kukreti, A., Painolli, A.K., Rana, N., Where do we stand: factors affecting sustainable development, Procedia Environmental Science, Engineering and Management, Volume 8, Issue 2, 2021, Pages 583-589

[24] Kulkarni, S & Chima, P 2021 'Challenges faced by UK university students due to thcoronavirus crisis in the Higher Education' Preprints.org. https://doi.org/10.20944/preprints202102.0192.v1

[25] Kulkarni, S & Sodre, JR 2017, An Assessment of Lightweight Vehicles to Reduce Greenhouse Gas Emissions with Focus on Magnesium. in Thoughts and Refections on the use of Magnesium. Birmingham City University, pp. 84-91.

[26] Kulkarni, S, Chapman, C & Shah, H 2016 'Computational Fluid Dynamics (CFD) Mesh Independency Study of A Straight Blade Horizontal Axis Tidal Turbine' Preprints.org. https://doi.org/10.20944/preprints201608.0008.v1

[27] Kulkarni, S, Chapman, C, Shah, H & Edwards, D 2018, 'A computational design method for bio-mimicked horizontal axis tidal turbines', International Journal of Building Pathology and Adaptation, vol. 36, no. 2, pp. 188-209. https://doi.org/10.1108/IJBPA-06-2017-0029

[28] Kulkarni, S, Chapman, C, Shah, H, Parn, E & Edwards, DJ 2018, 'Designing an efficient tidal turbine blade through bio-mimicry: A systematic review', Journal of Engineering, Design and Technology, vol. 16, no. 1, pp. 101-124. https://doi.org/10.1108/JEDT-08-2017-0077

[29] Kulkarni, S, Chapman, C, Shah, H, Parn, E & Edwards, DJ 2018, 'Design Study of Horizontal Axis Tidal Turbine Blade', Mindanao Journal of Science and Technology. https://www.open-access.bcu.ac.uk/10089/

[30] Kulkarni, S, Edwards, DJ, Chapman, C, Hosseini, MR & Owusu-Manu, D-G 2019, 'A preliminary mechanical design evaluation of the Wikispeed car: for light-weighting implications', Journal of Engineering, Design and Technology, vol. 17, no. 1, pp. 230-249. https://doi.org/10.1108/JEDT-09-2018-0154

[31] Kulkarni, S, Edwards, DJ, Parn, E, Chapman, C, Aigbavboa, CO & Cornish, R 2018, 'Evaluation Of Vehicle Lightweighting To Reduce Greenhouse Gas Emissions With Focus On Magnesium Substitution', Journal of Engineering, Design and Technology, vol. 16, no. 6, pp. 869-888. https://doi.org/10.1108/JEDT-03-2018-0042

[32] Kulkarni, SS, Wang, L, Golsby, N & Lander, M 2021, 'Fluid-structure interaction based optimisation in tidal turbines: A perspective review', Journal of Ocean Engineering and Science. https://doi.org/10.1016/j.joes.2021.09.017

[33] Malik, A., Kumar, A., Kandpal, H., Morphometric analysis and prioritization of sub-watersheds in a hilly watershed using weighted sum approach, Arabian Journal of Geosciencesthis link is disabled, 2019, 12(4), 118

[34] Mohanakurup, V.K., Baghela, V.S., Kumar, S., Srivastava, P.K., Doohan, N.V., Soni, M., Awa, H., 5G Cognitive Radio Networks Using Reliable Hybrid Deep Learning Based on Spectrum Sensing, Green Internet of Things in 5G and Beyond, Hindawi, Volume 2022

[35] Mohta, M., Soni, N., Choudhury, T., Kumar, V., Market Analysis on Li-Fi Technology, Proceedings of the 2019 8th International Conference on System Modeling and Advancement in Research Trends, SMART 2019, 2020, pp. 139–143, 9117427

[36] Nagwan, H., Rana, N., Kumar, A., Tewari, P.C., Performance and emission characteristics of diesel engine using Karanja oil methyl ester and Eucalyptus oil fuel blends, Procedia Environmental Science, Engineering and Management, Volume 6, Issue 4, 2019, Pages 553-563

[37] Painoli, A.K., Bansal, R., Singh, R., Kukreti, A., Impact of digital marketing on the buying behavior of youth with special reference to Uttarakhand state, Big Data Analytics for Improved Accuracy, Efficiency, and Decision Making in Digital Marketing, 2021, pp. 162–182

[38] Parag, J. (2011). Hybrid mixed handover for call blocking probability analysis in wireless ATM network. Journal of Engineering and Computer Innovations, 2(4), 68-77.

[39] Sharma, C., Singh, M.P., Chaudhary, S., Cost Analysis of Solar-Wind Based Hybrid Renewable Energy System, Lecture Notes in Electrical Engineeringthis link is disabled, 2021, 700, pp. 2733–2746

[40] Sharma, S., Chandra, U., & Jain, P. (2017). A literature survey on automation of test data generation for branch coverage testing using genetic algorithm. International Journal of Computational Intelligence Research, 13(6), 1521-1531.

[41] Varshney, R., Jain, P., & Vijay, S. (2018, September). Massive MIMO Systems In Wireless Communication. In 2018 2nd International Conference on Micro-Electronics and Telecommunication Engineering (ICMETE) (pp. 39-44). IEEE.

Chapter - 12

Machine Learning and AI (Health Care)

Dr. Anil W. Kale

Mahatma Gandhi Mission's College of Engineering and Technology, India. Email: anil5474@gmail.com

Dr. Nandkishor P Karlekar

MGMCET Kamothe Navimumbai, India. Email: npkarlekar.mgmcet@gmail.com

Dr. Vaibhav Narawade

Ramrao Adik Institute of Technology, India. Email:vnarawade@gmail.com

Dr. Sanjay B. Waykar

Mahatma Gandhi Mission's College of Engineering and Technology, India. Email:sbwaykar@gmail.com

ABSTRACT

Exabyte of pet bytes (or Exabyte of gigabytes) scale information to open up in various organizations and measurements. Given the complex and time-touchy wellbeing executions, 5G gadgets need broad finish to-to-end, edge, or highlight to-point delay, and different components, or attributes to be streamlined. the improvement of new advancements Device-explicit upgrades of 5G systems administration being right now are being acquainted with tackle different wellbeing explicit innovations like the Internet of Things (IoT) to give profoundly customized medicines, while the more extensive arrangement of wise organizations regularly tries to interface unique populaces like the IoT organizations and IoT gadgets for pervasive medical services applications that request explicit quality and expanding cell and region infiltration. Inside the setting of this examination, we talk about current 5G IoT-related issues like versatility, effectiveness, execution, openness, gadget and advancement, we tackle less energy use, adaptability, gadget accessibility, and battery life, alongside more-permitting gadgets to be gotten to, ease of use, and insurance are settled. Under AI, we have been attempting to grow the current admonition so such analytic medical services substances and way of life propensities can be represented and find more productive future therapies for by exploring data and automatable use situations identified with these elements.

BACKGROUND

Symptomatic instruments further develop the general medical care by conveying basic data to the appropriate people in real conditions. The general objective of powerful medical care is to make it simpler for the patient to get information about their findings and the best strategy for settling them. Through understanding the numerous parts of medical care, patients are engaged to react properly in case of intense disease. It urges medical services experts to broaden their practices by allowing a more worldwide far off turn upward to guarantee investment funds and empowers more savvy therapies for their patients through permitting the utilization of more current mechanical developments; a far reaching medical services foundation is expected to guarantee individuals have the wellbeing therapy they need. Aside from actual wellness, one of the basic objective is expanding medical care uses is to a push ahead by-regardless of step early analysis. As another model, the IoT area, this is anticipated to be worth 158.1 billion out of 2022, individual medical care the executives with the Internet of Things.

The Internet of Things will prompt a progress in medical care and lower the expense of clinical gear. IoT will acquire considerable use because of the sending would happen on account of 5G organizations. 5G organizations are set to be progressively well known for emergency clinics, clinical offices, and medical care suppliers. This is the fifth influx of versatile organization (5G). There are three ages of organizations following the primary 2G, 3G, 3G, and 4G organization. Another sort of organization has been fostered that interfaces basically all and everything to each other, making it conceivable to consolidate the mechanical and the actual domains. Regularly called multi Gig hertz remote innovation, which is promoted as having the option to give the Gbps, ultra-hertz transfer speed, colossal force, and consistency for customers, just as generous and generally development in speed for the organization utilization for clients. Expanded creation and expanded usefulness empower new industry areas to do what they already couldn't and help them to do all alone.

Applications that incorporate availability like public security and government utilities, and transportation, just as brilliant gadget portability for clinical and wellbeing checking, and enormous information are on 5G through and the colossal IoT. Albeit 5G is worked to have the ability to give different organizations over the long haul, it should be extended to be made adaptable to oblige the administrations that we don't yet perceive.

Upgraded portable broadband

As well as making our more brilliant telephones more productive, 5G will empower an expansion, all things considered, inundation encounters like VR and AR, higher information speeds, and less-per-second execution, it will take into account a decline in dormancy.

Those fundamental for achieving the mission

No new help has arisen as an industry relies so vigorously upon high on super solid, broadly open, low-idleness associations as doing as such with the 5G-empowered arrangement.

Quickly growing Internet of Things

The adaptability and versatility of 5G innovation was expected to assist with diminishing information speeds, asset use, and openness costs for almost whatever is introduced in a wide scope of articles in the organization, taking into consideration exceptionally productive and adaptable, however savvy, arrangements. With the multiplication of remote foundation, brilliant regulators, and keen gadgets, various enterprises are depending on disseminated machine-to-machine (M2M) conditions and web of things (IoT) for working on functional productivity.

Existing M2M organizations require various regulators and heterogeneous equipment/correspondence interfaces to interface the M2M sensors to their particular focal units to get measure and send information. In any actual climate, the sent M2M/IoT stage gathers organized and unstructured information and an assortment of related and inconsequential never-ending occasions which are handled by at least one hubs relying upon its physical and useful capacities.

The regulators and interfacing M2M gadgets generally have pre-taken care of programming that takes into account determined circumstances and where the runtime works on a proper arrangement of rules/rationale inside its limit of tasks. Because of the imperatives of preparing and correspondence limit, any single hub with preloaded underlying knowledge will have restricted extension to associate and total crude occasions, create situational mindfulness and make brief moves to deal with the problematic functional climate of a M2M/IoT environment. These frameworks have restricted capacity to use machine knowledge and cooperative learning inside a predefined framework limit. The cognizance interaction in these frameworks is for the most part bound to information mining and in view of restricted pre-taken care of knowledge.

As specialists have done in the course of the most recent couple of years, record and patient's experience data is held for a few ages. With patients who decide to utilize the investigation, the records would have been kept in an office place so they may see it. the issue with communicating clinical notes to the primary second anybody sees a patient to another specialist is that it is a mix-up Once the patient's records had been shipped off the new doctor, which will be accomplished through reflecting and movement. Lamentably, duplicates can't be delivered and at this point don't move, on the grounds that there would be nothing else of the asset.

It has been set up that electronic checking for medical services suppliers and their patients to assist individuals with bettering screen their wellbeing and cling to therapy plans It is regularly when medical services associations apply such predictable practices, are in any event, while moving them to another association, frequently unintentionally, that information might be ported.

Normally, there are a few classified records situated in workplaces that are private, including emergency clinics, army installations, and outsider safety net provider offices that can't be found by the patient's present specialist or advisor. Albeit this data is unquestionably essential to patients, all things considered, it very well might be scattered to the doctors, centers, and everybody in a question of crisis as I in a period of critical clinical scarcity, for I have thirty wellbeing organizations to work with, there isn't an approach to send it to the overall population and to such gatherings as that as effectively as this to the current establishments.

As this assignment of making clinical data all the more broadly accessible has been cultivated, admittance to it will be altogether streamlined. Notwithstanding all the should be distant from everyone else, being overlooked, unfortunately, permitting autonomous individuals the ordinary chance to utilize the innovation, just as protected people to see clinical records is an issue we actually need to confront. Patients may be in a difficult situation when endeavoring to track down their clinical data, on the grounds that the records are generally put away in a facility, and not in every case close by.

However, by and large, the vast majority of the records stay at the center or medical clinic where a patient's fundamental doctor or attendant specialist is found. The people who travel excessively far from their center for treatment is by all accounts unconscious of this material, nonetheless, regularly track down that the greater part of it is on record or not carefully accessible.

OBJECTIVES

- The target of the innovation is to a doing is moving from a conventional expert and emergency unit a scattered clinic, and there are numerous advances in transit that are empowering this transformation of "from the base up". The machine's innovation starting today, 5G and other systems administration gadgets are in medical care, 5G frameworks are generally being used for getting to medical services offices of the learned person, and patient administration advancements are headed to advance.
- The other target of the innovation is to a Due to the expected expansion in the volume of information created by present day approaches like modernized medical services, virtual medical care, and other required applications, we ought to expect a gigantic piles of Exabyte of pet bytes (or Exabyte of gigabytes) scale information to open up in various configurations and measurements. Given the complex and time-delicate wellbeing executions, 5G gadgets need broad finish to-to-end, edge, or highlight to-point delay, and different components, or attributes to be streamlined.
- The other target of the creation is to a the improvement of new advances Device-explicit upgrades of 5G systems administration being as of now are being acquainted with tackle different wellbeing explicit advances like the Internet of Things (IoT) to give exceptionally customized medicines, while the more extensive sending of savvy networks regularly tries to connect unique populaces like the IoT organizations and IoT gadgets for omnipresent medical services applications that request explicit quality and expanding cell and region infiltration.
- The other target of the innovation is to a Within the setting of this investigation, we talk about current 5G IoT-related issues like adaptability, productivity, execution, openness, gadget and streamlining, we tackle less energy use, versatility, gadget accessibility, and battery life, alongside more-permitting gadgets to be gotten to, convenience, and insurance are settled. Under AI, we have been attempting to grow the current admonition with the goal that such analytic medical services elements and way

of life propensities can be represented and find more proficient future therapies for by investigating data and automatable use situations identified with these elements.

SUMMARY

A short synopsis of parts and explicit subtleties of the capacity to follow: (A primer definition or not many of many complete ideas canvassed in the extensive extra text) the text is organized so as to exclude a serious investigation of anything in the psyche, however rather centers around different objectives and thought processes. The ensuing audit isn't planned to contain something past of the scope of examination, nor is it expected to have any sort of effect.

As indicated by an excellent element, the straightforwardness, the structure is described by a capacity for substantial and constant or close fundamental choice help for incorporated M2M and IoT applications. The intellectual insight stage has a one of a kind equipment, programming, and correspondence plan that reproduces the underlying and functional model of a concentrated sensory system or neural organization of vertebrates, like the human sensory system, through common versatile communitarian learning, versatile information sharing, continuing and applying steady information logically, and self-change through common sharing of liabilities.

It depends on a unique run-time that can resolve the setting of a changing physical and functional climate through communitarian activities which keep on developing over the long haul. The framework can self-change the whole biological system to adjust to changing conditions and serve constant functional effectiveness.

As indicated by another excellent perspective, the stage configuration assigns functional abilities, conditions and progressions to deal with occasions, execute tasks and convey upholds. The intellectual insight stage has three legitimate handling layers—edge regulators (EC), terminal aggregators (TA), and a wise assistance the executives center (ISM or Hub). The edge regulators go about as receptors for explicit activities. They recover the crude occasions from the actual climate and broadcast them to friends and terminal aggregators to help setting explicit activities.

The terminal aggregators direct and support the activity control and information dissemination for edge regulators, and work with cooperation in or close to ongoing across the layers on a level plane and in an upward direction. They are associated with different TAs and ECs for data trade. They are additionally associated with the ISM for trade of orders and functional information. The Intelligent Service Management Hub screens activities and utilizations complex measurable and AI strategies and calculations to determine situational needs and update rules for subsystems on the fly. The Hub might be sent in at least one workers.

As per one part of the exposure, the unique run-time conduct of the revealed stage is cultivated in two stages—every hub has an essential degree of insight and rules to handle content, information, occasions, and work together with different hubs. Also, the pecking order of layers permits the stage to total the learning at each even out and change the obligations. The unveiled intellectual stage decouples the principles from the processors, the undertakings from the work process subroutines/specialists, and agents the preparing of the knowledge into the coherent layers to have independence in choice help and adjust changes consequently.

The specialists can rearrange themselves inside the runtime climate to work on functional independence across the layers. The specialists might get across the preparing layers depending on the situation relying upon the capacities of the specialist and the prerequisites of the framework. The development or organization of the specialists is worked with by the runtime to work on the continuous execution. The development or sending of the specialists might be dynamic or manual.

Every one of the layers have uniform equipment and programming design so the framework can convey the specialists, update rule motors, and interpret the changing runtime semantics into an extremely durable functional climate. All hubs, including regulators, aggregators and ISM workers, are preloaded with least programming (specialists, rules motor, and occasion transport structures) and the ISM has a vault of specialists. Inside the associated climate, every specialist sends ongoing notices to its friends and manager layers to look for help to deal with exemption conditions.

The reaction from the supporting hubs is a prompt activity or potentially change in the handling rules or learning examples to help the conditions just as conveying any extra specialists to follow up on those circumstances. The stage engineering upholds the framework to characterize the need and request of such changes.

As indicated by one more part of the divulgence, the unveiled intellectual stage has a collective runtime climate where every hub is a member. This backings common learning and sharing and results in new information for the framework. This permits the framework to survey, resolve and cycle the specific circumstance and content progressively and convey changed activities and administrations in segregation and additionally as a team with different hubs and give continuous functional knowledge. Consequently, the conveyed stage develops the information on the stage and helps the framework acquire independence by improving situational mindfulness and decentralized dynamic capacities over the long run.

As given in the methodology and program as per one illustration of the current advancement, techniques and frameworks are accommodated permitting the patient to give or overseeing medical services specialist admittance to the patient's clinical record. Another valuable, smaller, straightforward, and easy to utilize highlight controlled gadget for putting away, overseeing, and putting away wellbeing records can be given.

As per another understanding, there is a framework, there is a wellspring of clinical data, a beginning of clinical information for the customer, just as at a data set for which man-made consciousness (AI) included; then, at that point either or a greater amount of these substances is associated with each method for delight information utilizing a man-made reasoning (AI) part, which is utilized to break down the individual information, perceive no less than one worry that should be dealt with by the doctor, and either made development on by an autonomous occasion, or by an element allowed authority given to the patient. data that distinguishes something like one client, unquestionably somewhat through a patient's name or extraordinary identifier may be made accessible to the next This framework is outfitted with essentially with one accessible ongoing, and without a doubt another association that utilizes the Health Level 7 convention.

One may take a gander at the extension part, the annoyance hypothesis component, or both: the articulation and the definition. the basic suspicions of AI historical background can likewise include a symptomatic convention; a treatment convention; and a derivation motor. It is conceivable that the AI computation part might create the methodology for acquiring the treatment and/ seeking after the treatment from somewhere around one clinical wellspring of information.

Somewhere around one of a reasonable, conceivable, presumably life, clinically huge, or significant medication communications, a plausible impact, drug poisonousness, or realized hypersensitivity was recognized; preceding a patient appraisal, ID of an issue, there are probably going to be potential or conceivable huge effects; prior to arriving at a decision on clinical therapy, one should seriously mull over if present; prior to leading a clinical examination, a clinical trial was proposed by the AI; and control medicine by the chance of harmfulness was distinguished by the clinical AI; and during clinical assessment, suggestions, doctors proposed a test was made by patients for medicines or for drug tests.

Somewhere around one of the sorts of the AI instruments that the specialist may envision is: fluffy rationale, which is frequently viewed as the delicate; man-made consciousness (regularly alluded to as delicate or machine); a degree of information (a mechanized type of a book of information, a. especially fake information); a choice emotionally supportive network (particularly fluffy rationale based, softbot, or some type of programming computerization). It can additionally give an information motor to coordinate, document, alter, save, organization, and interaction clinical information from the individual information and the information.

Moreover, there is something like one: An essential kind of clinical aide; A continuous program that is done by the medical services administrative board that is determined by rules called a Condition; or rule that are ordinarily used to treat or help a clinical issue that is utilized as one sort of an administrative board wellbeing; Or by numerous other wellbeing guides, the guideline framework utilizes Codes like ICD-9 and CPT-4 for instance.

Both of the three wellsprings of individual clinical information might be at least one living will, including explicit medication or transcribed solutions, to which the other, share genuine, or video-or medication record remedies, including Wills are sent before an advancement, either scripts ready before Advancement of Will or protection commitment, or one penmanship notes, or dental or heart biopsies with composed requests, and pre-coded client results, if any of those three are sent, no endeavor will be made "to make a big difference forever", or venture down (coding orders that advise the clinical staff to don't endeavor to save patients) will be started.

It is very conceivable that the monetary and private information are being accessible to: the at generally one of the patient to convey; The monetary and private information are broadly appropriated, which implies that they may have been duplicated and reallocated: there is a huge danger that the two people will get individual and clinical information while interfacing with just one of the entirety of getting the hang of, protecting, getting to, investigating, joining, or altering is warning, reaching, planning, or person to person communication information.

DIAGRAM

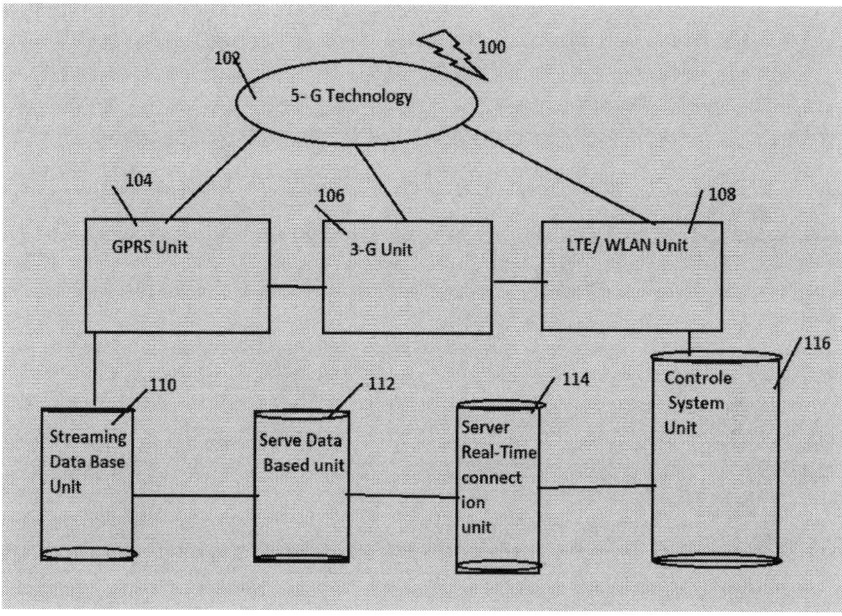

Fig. 1: Technologies towards 5G Network Block Diagram.

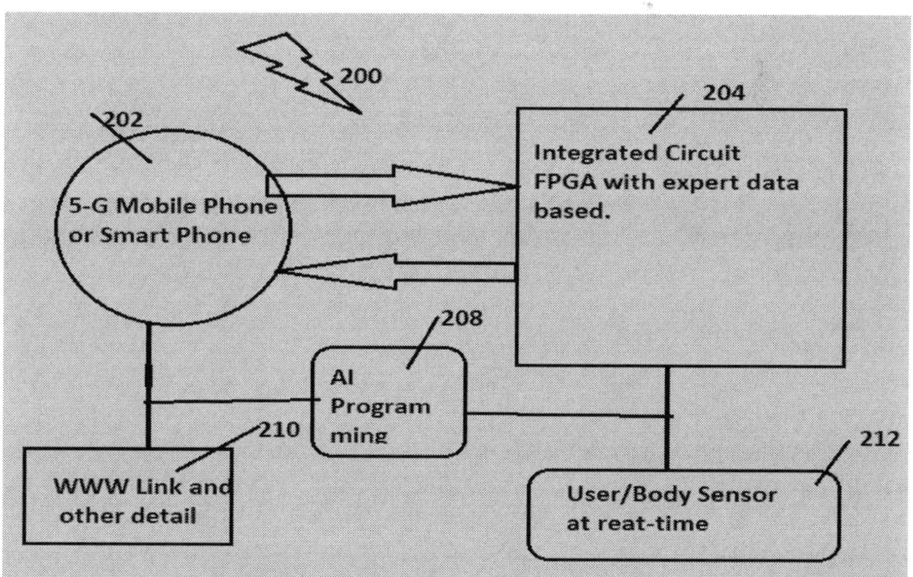

Fig. 2: Health-Care block diagram.

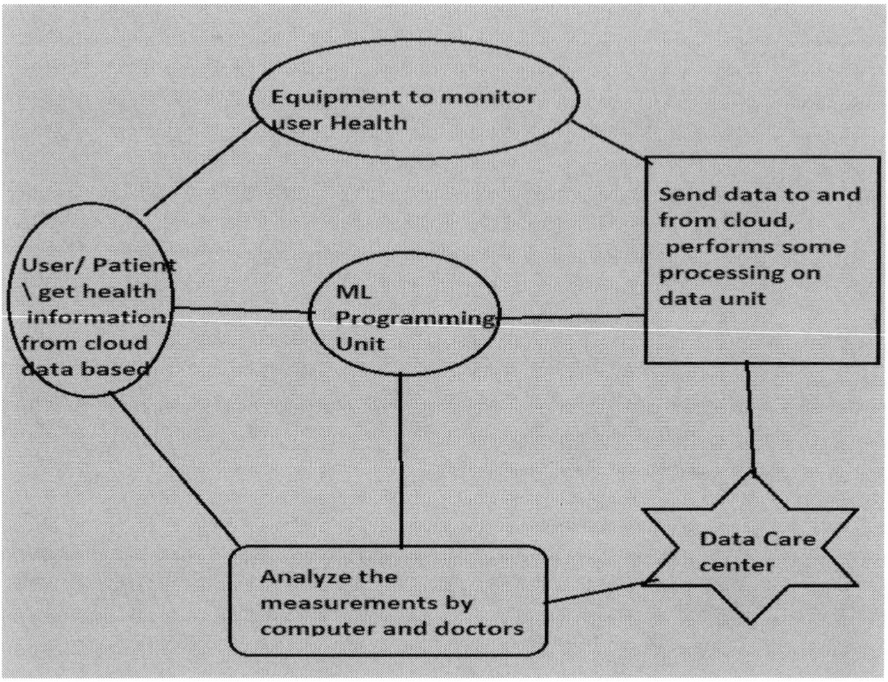

Fig. 3: Health-Care Using IoT notification with Machine Learning Programming.

DESCRIPTION

Fig. 1: Presently, there are very few data transmissions on the lookout; subsequently, 5G organizations would need to give a wealth of content comparative with some days. They would likewise need to help heterogeneous traffic, for example, machine-to-to-machine (or gadget) machine made by an expanding measure of Internet of Internet of Things gadgets. Traffic request might be diverse in both reality and in degrees. While in this work, we fight that it is feasible to reach both of these objectives across the organization; it is additionally astute to focus on adaptability and configurability.

The organization framework that can uphold these requests is alluded to as a "Flex5G" by us. In the other hand, we look at associations among adaptability and configurability, and how the on-the-the-fly MAC modification and organization assets arranging take into account foundation and content information to collaborate. The discoveries we found, both through prototyping and in reenactments, showed that network conventions do overall don't associate with one another at a significant issue for exceptionally constructive outcome, and we additionally saw considerable improvement in getting to subnet explicit subnets with diminished idleness.

We close with an examination of what the Flex5G stage plans would mean for expected normalization and present a portion of the featured components concerning a couple significant 5G systems administration facilitators.

Fig. 2: Health-Care block chart. is just about as fundamental as staying away from sickness (anticipation) This includes turning out to be truly dynamic, assurance against measles, safe water, and not burning-through liquor or utilizing cigarettes. Without a doubt, keeping the home and environmental factors clean adds to a solid body and psyche. We may have the best treatment, however that would have little effect if the actual instrument was not working as expected. Moreover, that direct actual upgrades to one may utilize programs, one might utilize individual upkeep strategies to support the overall wellness and well-enhancementing them too. Despite the fact that actually keep up with great wellbeing, the medical services area should in any case enjoy preventive proportions of sickness.

Medical problems, yet frequently a compromised social prosperity is given to a been permitted to youngster put on weight, as higher possibilities of creating diabetes and all of getting dismissed for exercises on account of the extra weight it puts on the social circle. Furthermore, financing of society in this field is a significant issue that we should fight with, as seen with the crime pointed toward smothering smoking are ill-conceived.

The capacity to distinguish/ability to recognize some medical problems

In the event that people may not report the data, it will make it harder for the wellbeing office to help them. Significant and early recognizable proof and effort from the medical care local area was pivotal. Disease screening is a basic and informational model If the quickly developing melanoma isn't being distinguished, the anticipation and screening cycles may not be extremely effective.

They would be bound to focus on their own government assistance in case they were partaking in wellbeing and health checks. Longitudinal insights are more significant in case they are acquired consistently. Realizing how hypertension or the amount you are improving outcomes consistently is a higher priority than simply gathering this information once a couple of times each year.

Deciding the idea of infections

We should propel this treatment to this next stage to have some effect. Maybe costlier is getting the outcome inaccurately, by utilizing some unacceptable treatment. Idealness in discovery is pivotal for the beginning of specific infections. The medical care framework gives individuals with the aptitude and methodology for these exercises.

Physiotherapy treats illnesses

One of the most basic issues of medical services the present exchange is the ability to both fix and help people who endures." One way for residents to join forces with the emergency clinics is to tell them their musings and sentiments by utilizing the right to self-administration. Managing issues that might emerge during the patient's last days

A difficult yet life-preserving method can well be fruitful, however on the off chance that its productivity drops, it has a more regrettable result and if a medication stops a patient's consideration during the fitting season of most extreme advantage, far from death, there is the conceivable downside of making personal satisfaction issues during the patient's last range of life. Then again, sicknesses that create due to age, like some more slow maturing illnesses, prompt more debasement of some human frameworks, and some speedier.

The clinical ranges of abilities are characterized by side effects or organ: the heart, lungs, eyes, mind, and different afflictions have separate divisions (i.e., respiratory, psychological wellness issues, and so forth) untimely withdrawal from treatment in a coma could be pretty much as hurtful as, if not more in this way, to the individual who is really focused on as an exhaustive a view as an all-encompassing need exists, there would be a considerable ascent in the quantity of experts managing the older populace.

Fig. 3: The advantage of the use of IoT advancements gives clinical experts and patients admittance to the most recent cutting-edge clinical hardware. Regardless of the benefit of checking and computerizing the current environment, AI might help with groupings like synthetic and mechanical frameworks of differing degrees. The essential goal for medical care is to tackle the issue of bringing down the cost and further developing the patients' well-level of the framework, which are perspectives that are both important for business applications that utilization the IoT.

There is a proceeding with need for the most current executions in the medical services world. Scholarly or effective fixes are needed to resolve gives that emerge with IoT clinical consideration is investigated, and right now under assessment. In particular, concerning different sorts of implantable and removable or smaller gadgets for chip, and IoT frameworks, analyzed in deciding the information transmission Where there is a requirement for an implantable part, the implantation of a substitution would be a typical event. While making a wearable's organizations have so far beat a few of their impediments, this investigation shows a couple all the more plainly show their excess obstacles. The issues and concerns found in the medical coverage market were tended to in this article, which for the most part managed Machine Learning (ML) and the Internet of Things (IoT) use in the medical care industry.

The examples of Machine Learning in IoT utilizes calculations can be shown by showing the different strategies utilized in their turn of events, assembled per the cycle, and masterminded arranged by use, and recognized from others by featuring the exploration papers where they were first distributed.

OUTCOME

1. One striking movement that we're anticipating doing is moving from a conventional expert and emergency unit a scattered medical clinic, and there are numerous advancements in transit that are empowering this upset of "from the base up". The machine's creation starting today, 5G and other systems administration gadgets are in medical services, 5G frameworks are generally being used for getting to medical care offices of the intelligent person, and patient administration advances are en route to advance. Because of the expected expansion in the volume of information produced by current methodologies like mechanized medical services, virtual medical care, and other required applications, we ought to expect a colossal piles of Exabyte of pet bytes (or Exabyte of gigabytes) scale information to open up in various configurations and measurements. Given the complex and time-delicate wellbeing executions, 5G gadgets need broad finish to-to-end, edge, or highlight to-point delay, and different elements, or qualities to be advanced. the improvement of new advances Device-explicit upgrades of 5G systems administration being right now are being acquainted with tackle different wellbeing explicit innovations like the Internet of Things (IoT) to give exceptionally customized medicines, while the more extensive organization of keen organizations frequently looks to connect dissimilar populaces like the IoT organizations and IoT gadgets for universal medical services applications

that request explicit quality and expanding cell and region infiltration. Inside the setting of this examination, we talk about current 5G IoT-related issues like versatility, productivity, execution, openness, gadget and advancement, we tackle less energy use, adaptability, gadget accessibility, and battery life, alongside more-permitting gadgets to be gotten to, convenience, and security are settled. Under AI, we have been attempting to extend the current admonition so such symptomatic medical care substances and way of life propensities can be represented and find more effective future therapies for by surveying data and automatable use situations identified with these elements.

2. According to claim1# the creation is to a One eminent shift that we're anticipating doing is moving from a conventional expert and emergency unit a scattered clinic, and there are numerous advancements in transit that are empowering this transformation of "from the base up". The machine's innovation starting today, 5G and other systems administration gadgets are in medical care, 5G frameworks are for the most part being used for getting to medical services offices of the erudite person, and patient administration advancements are headed to advance.
3. According to claim1# the creation is to a Due to the expected expansion in the volume of information produced by current methodologies like mechanized medical services, virtual medical services, and other required applications, we ought to expect a colossal piles of Exabyte of pct bytes (or Exabyte of gigabytes) scale information to open up in various arrangements and measurements. Given the complex and time-delicate wellbeing executions, 5G gadgets need broad finish to-to-end, edge, or highlight to-point delay, and different provisions, or attributes to be enhanced. the improvement of new innovations Device-explicit upgrades of 5G systems administration being at present are being acquainted with tackle different wellbeing explicit advances like the Internet of Things (IoT) to give profoundly customized medicines, while the more extensive sending of keen organizations frequently looks to interface dissimilar populaces like the IoT organizations and IoT gadgets for universal medical care applications that request explicit quality and expanding cell and region entrance.
4. According to claim1# the creation is to a Within the setting of this examination, we talk about current 5G IoT-related issues like adaptability, proficiency, execution, openness, gadget and streamlining, we tackle less energy use, versatility, gadget

accessibility, and battery life, alongside more-permitting gadgets to be gotten to, ease of use, and assurance are settled. Under AI, we have been attempting to extend the current admonition so such indicative medical care substances and way of life propensities can be represented and find more productive future therapies for by checking on data and automatable use situations identified with these elements.

REFERENCES

[1] Abhigna, B.S., Chakraborty, S., Kumar, V., Medha: A collaborative E-concept framework, Proceedings of the 2018 International Conference On Communication, Computing and Internet of Things, IC3IoT 2018, 2019, pp. 117–122, 8668127

[2] Abidin, S., Vadi, V.K., Rana, A., On Confidentiality, Integrity, Authenticity, and Freshness (CIAF) in WSN, Advances in Computer, Communication and Computational Sciences, 2020, pp 87-97

[3] Ahay, A., Choudhury, T., Kumar, V., Hierarchical Evolution of Digital Games, Proceedings of 2019 International Conference on Computational Intelligence and Knowledge Economy, ICCIKE 2019, 2019, pp. 381–386, 9004378

[4] Area: Communication, Frequency modulation, Networks, wireless communication

[5] Area: Communication, Frequency modulation, Networks, wireless communication

[6] Area: Communication, Frequency modulation, Networks, wireless communication

[7] Area: Communication, Frequency modulation, Networks, wireless communication

[8] Arora, M., Kumar, V., Dave, M., Hybrid cloud infrastructure for load balancing and scaling with vm path optimization using ant colony optimization, Journal of Advanced Research in Dynamical and Control Systems, 2020, 12(5 Special Issue), pp. 660–674

[9] Arora, M., Kumar, V., Dave, M., Task scheduling in cloud infrastructure using optimization technique genetic algorithm, Proceedings of the World Conference on Smart Trends in Systems, Security and Sustainability, WS4 2020, 2020, pp. 788–793, 9210303

[10] Bansal, R., Gupta, A., Singh, R., Nassa, V.K., Role and impact of digital technologies in E-learning amidst COVID-19 pandemic, Proceedings - 2021 4th International Conference on Computational Intelligence and Communication Technologies, CCICT 2021, 2021, pp. 194–202

[11] Bansal, R., Obaid, A.J., Gupta, A., Singh, R., Pramanik, S., Impact of Big Data on Digital Transformation in 5G Era, Journal of Physics: Conference Seriesthis link is disabled, 2021, 1963(1), 012170

[12] Chauhan, S.S., Khare, A.K., Kinematic analysis of the ABB IRB 1520 industrial robot using roboanalyzer software, Evergreenthis link is disabled, 2020, 7(4), pp. 510–518

[13] Chauhan, S.S., Khare, A.K., Method of Obtaining Planar State of Stress Using Mohr's Circle—Some Typical Cases, Lecture Notes in Mechanical Engineering, 2022, pp. 33–38

[14] Goel, S., Tripathi, A., Choudhury, T., Kumar, V. Parkinson's Disease Detection using Soft Computing Technique, Proceedings of the 2019 8th International Conference on System Modeling and Advancement in Research Trends, SMART 2019, 2020, pp. 3–6, 9117384

[15] Govindarajan, V., Sivakumar, R., Patil, P.P., Kannan, M., Ramesh, B., Effect of Tungsten Carbide Addition on the Microstructure and Mechanical Behavior of Titanium Matrix Developed by Powder Metallurgy Route, Advances in Materials Science and Engineeringthis link is disabled, 2022, 2022, 2266951

[16] Gupta, A., Singh, R., Nassa, V.K., Sharma, P., Koti, K., Investigating Application and Challenges of Big Data Analytics with Clustering, 2021 International Conference on Advancements in Electrical, Electronics, Communication, Computing and Automation, ICAECA 2021, 2021

[17] Jain, P., Vijay, S., & Gupta, S. C. (2009). ATM Network QoS Parameters and Performance Analysis using OPNET Simulator. International Journal of Recent Trends in Engineering, 2(4), 228.

[18] Jain, P., Vijay, S., & Gupta, S. C. Performance Analysis & QoS Guarantee in ATM Networks. Global Journal of computer Science and Technology, 131-136.

[19] Jain, S., Choudhury, T., Kumar, V., Kumar, P., Detecting malware and analysing using sandbox evasion, Proceedings of the 2018 International Conference On Communication, Computing and Internet of Things, IC3IoT 2018, 2019, pp. 111–116, 8668149

[20] Kala, S., Nandan, H., Sharma, P., Shadow and weak gravitational lensing of a rotating regular black hole in a non-minimally coupled Einstein-Yang-Mills theory in the presence of plasma, European Physical Journal Plusthis link is disabled, 2022, 137(4), 457

[21] Kandpal, H., Kumar, A., Malik, A., Morphometric analysis and prioritization of sub-watersheds in Bino watershed, Uttarakhand: A remote sensing and GIS perspective, Indian Journal of Ecology, 2018, 45(3), pp. 475–478

[22] Kannan, M., Singh, S., Prasad, R.R., Synthetic Methods to Obtain Calcia-Stabilized Zirconia Powders—A Review, Lecture Notes in Mechanical Engineeringthis link is disabled, 2021, pp. 405–416

[23] Kukreti, A., Painolli, A.K., Rana, N., Where do we stand: factors affecting sustainable development, Procedia Environmental Science, Engineering and Management, Volume 8, Issue 2, 2021, Pages 583-589

[24] Kulkarni, S & Chima, P 2021 'Challenges faced by UK university students due to thcoronavirus crisis in the Higher Education' Preprints.org. https://doi.org/10.20944/preprints202102.0192.v1

[25] Kulkarni, S & Sodre, JR 2017, An Assessment of Lightweight Vehicles to Reduce Greenhouse Gas Emissions with Focus on Magnesium. in Thoughts and Refections on the use of Magnesium. Birmingham City University, pp. 84-91.

[26] Kulkarni, S, Chapman, C & Shah, H 2016 'Computational Fluid Dynamics (CFD) Mesh Independency Study of A Straight Blade Horizontal Axis Tidal Turbine' Preprints.org. https://doi.org/10.20944/preprints201608.0008.v1

[27] Kulkarni, S, Chapman, C, Shah, H & Edwards, D 2018, 'A computational design method for bio-mimicked horizontal axis tidal turbines', International Journal of Building Pathology and Adaptation, vol. 36, no. 2, pp. 188-209. https://doi.org/10.1108/IJBPA-06-2017-0029

[28] Kulkarni, S, Chapman, C, Shah, H, Parn, E & Edwards, DJ 2018, 'Designing an efficient tidal turbine blade through bio-mimicry: A systematic review', Journal of Engineering, Design and Technology, vol. 16, no. 1, pp. 101-124. https://doi.org/10.1108/JEDT-08-2017-0077

[29] Kulkarni, S, Chapman, C, Shah, H, Parn, E & Edwards, DJ 2018, 'Design Study of Horizontal Axis Tidal Turbine Blade', Mindanao Journal of Science and Technology. https://www.open-access.bcu.ac.uk/10089/

[30] Kulkarni, S, Edwards, DJ, Chapman, C, Hosseini, MR & Owusu-Manu, D-G 2019, 'A preliminary mechanical design evaluation of the Wikispeed car: for light-weighting implications', Journal of Engineering, Design and Technology, vol. 17, no. 1, pp. 230-249. https://doi.org/10.1108/JEDT-09-2018-0154

[31] Kulkarni, S, Edwards, DJ, Parn, E, Chapman, C, Aigbavboa, CO & Cornish, R 2018, 'Evaluation Of Vehicle Lightweighting To Reduce Greenhouse Gas Emissions With Focus On Magnesium Substitution', Journal of Engineering, Design and Technology, vol. 16, no. 6, pp. 869-888. https://doi.org/10.1108/JEDT-03-2018-0042

[32] Kulkarni, SS, Wang, L, Golsby, N & Lander, M 2021, 'Fluid-structure interaction based optimisation in tidal turbines: A perspective review', Journal of Ocean Engineering and Science. https://doi.org/10.1016/j.joes.2021.09.017

[33] Malik, A., Kumar, A., Kandpal, H., Morphometric analysis and prioritization of sub-watersheds in a hilly watershed using weighted sum approach, Arabian Journal of Geosciencesthis link is disabled, 2019, 12(4), 118

[34] Mohanakurup, V.K., Baghela, V.S., Kumar, S., Srivastava, P.K., Doohan, N.V., Soni, M., Awa, H., 5G Cognitive Radio Networks Using Reliable Hybrid Deep Learning Based on Spectrum Sensing, Green Internet of Things in 5G and Beyond, Hindawi, Volume 2022

[35] Mohta, M., Soni, N., Choudhury, T., Kumar, V., Market Analysis on Li-Fi Technology, Proceedings of the 2019 8th International Conference on System Modeling and Advancement in Research Trends, SMART 2019, 2020, pp. 139–143, 9117427

[36] Nagwan, H., Rana, N., Kumar, A., Tewari, P.C., Performance and emission characteristics of diesel engine using Karanja oil methyl ester and Eucalyptus oil fuel blends, Procedia Environmental Science, Engineering and Management, Volume 6, Issue 4, 2019, Pages 553-563

[37] Painoli, A.K., Bansal, R., Singh, R., Kukreti, A., Impact of digital marketing on the buying behavior of youth with special reference to Uttarakhand state, Big Data Analytics for Improved Accuracy, Efficiency, and Decision Making in Digital Marketing, 2021, pp. 162–182

[38] Parag, J. (2011). Hybrid mixed handover for call blocking probability analysis in wireless ATM network. Journal of Engineering and Computer Innovations, 2(4), 68-77.

[39] Sharma, C., Singh, M.P., Chaudhary, S., Cost Analysis of Solar-Wind Based Hybrid Renewable Energy System, Lecture Notes in Electrical Engineeringthis link is disabled, 2021, 700, pp. 2733–2746

[40] Sharma, S., Chandra, U., & Jain, P. (2017). A literature survey on automation of test data generation for branch coverage testing using genetic algorithm. International Journal of Computational Intelligence Research, 13(6), 1521-1531.

[41] Varshney, R., Jain, P., & Vijay, S. (2018, September). Massive MIMO Systems In Wireless Communication. In 2018 2nd International Conference on Micro-Electronics and Telecommunication Engineering (ICMETE) (pp. 39-44). IEEE.

Chapter - 13

AI - Based Programming (Parking Light Monitoring)

Dr. Vaibhav Narawade

Ramrao Adik Institute of Technology, India. Email:vnarawade@gmail.com

Dr. Sanjay B. Waykar

Mahatma Gandhi Mission's College of Engineering and Technology, India. Email:sbwaykar@gmail.com

Dr. Anil W. Kale

Mahatma Gandhi Mission's College of Engineering and Technology, India. Email: anil5474@gmail.com,

Dr. Nandkishor P Karlekar

MGMCET Kamothe Navimumbai, India. Email: npkarlekar.mgmcet@gmail.com

ABSTRACT

Our development Intelligent Street and Parking Light Monitoring Through AI is a streetlamp shrewd checking and continuous information control and the executive's framework and philosophy for gathering sets of data from geologically conveyed areas worldwide, area, s or characterized sensor hubs committed to characterized streetlamps of a multi-light streetlamp framework. The development is energy utilization in metropolitan urban communities, little city, town and alternate way are developing step by step and in each city, a huge amount of power is being utilized for the possibility of road lighting another framework. It's obviously true that the force use by streetlamps in certain pieces of the city is equivalent to that of regions with an extreme focus transducer, yet it is noticed that the volume of energy utilized by streetlamp in these areas is something very similar. The advancement is in like manner a Street lighting is a fundamental element of wellbeing plans in houses and business establishments just as in outside parks, streetlamp the executives has a critical capacity in a brilliant observing plan that can be performed vivaciously in genuine checking conditions. In the PC share highlight different methodologies sent from the open organization PC from the Micro regulator. It works in two essential modes, the technique for actual guideline and directing. Through cooperative control mode, the lights are both turned on or on. As the day progressed, the lights are turned off in physically worked mode, and are empowered at dull. The organization is exceptionally effective in power preservation; however, it is fundamental that the framework mode is constantly improved. It would be more secure with the utilization of a strobing module. To improve the energy preservation of the streetlight, a worldly force control has been made. In a similar geological region, the stargazing clock is investigated for experimental impacts. The unit has a LED light splendour controller, an ongoing clock, and a MOSFET circuit. While energy saving sounds wonderful, the plan actually has mishaps. The MOSFET has high impedance and is equipped for causing static force misfortune. Also, the obstruction of MOSFET is extremely low. The frameworks watch the ordinary power design for one year and align the glow of the street against PWM as indicated by perceptions. This framework cap doesn't meet the indistinct propensity or inconstancy in the air.

BACKGROUND

Traffic stream may likewise be expanded by the arrangement of ongoing traffic refreshes for drivers. Frequently the neighbourhood stations have traffic reports. In any case, radio broadcasts don't have realities for the most part progressively.

Consequently, vehicles are at times caught in gridlocks preceding being educated by the radio broadcast about the case. Other than this a portion of the vehicles, in particularly those at provincial positions or heading in different ways, don't get precise traffic refreshes from neighbourhood radio broadcasts. Moreover, radio traffic reports are just for travelers going along extensions or streets and not ordinarily for explorers on more modest, nearby streets or roads in networks.

The absence of territorial insight on transportation disallows motor drivers from forestalling nearby sticks or contamination puts that the news stations don't communicate. Initially, upgraded traffic signal administration measures and subtleties on traffic to vehicles are required and needed progressively relying upon their position, and their course.

The development identifies with a framework for the observing of streetlamps whereby individual light disappointments are accounted for to a far off administration place that uses a heuristic remote correspondences organization or framework involving a majority of topographically divided or genuinely isolated hubs. The abilities of every one of the hubs incorporate closest sensor neighbor recognition and recording, programmed correspondence course assurance to a hub regulator and giving continuous status data to the hub regulator.

The hubs are fit for detecting, dissecting and executing the best accessible interchanges way to different hubs and to the hub regulator. The hub regulator itself might be a sensor with extra capacities, for example, correspondence capacity to a Management Center, commonly by remote means. The hub regulator additionally contains the essential rationale to play out the control hub capacities.

Lighting frameworks are accommodated roads, highways, spans, air terminals, oceanic ports, parking garages and government establishments for wellbeing and security purposes. The assurance of the working qualities of individual lights is normally done through a manual investigation. At the point when a light is observed to be non-working, a maintenance activity is directed on the light. Some more current lighting frameworks use light congregations that join an interior sensor to decide a light disappointment and a way to impart the disappointment condition.

For instance, the measure of energy being burned-through per unit time is a marker of a light's legitimate activity. Another methodology is simply the utilization of a photodiode that makes them genuinely take a look at hardware to decide right working with dusk or dawn. Many lighting frameworks in America don't utilize wise light gatherings that can detect and report their singular light functional status. This innovation gives a way to recognize and report a singular light's functional qualities without change to the current lighting framework electrical foundation.

The light sensor in this creation requires no alteration to the electrical framework of the current lighting framework. The light sensor searches energy from the climate to work. The sensor is precisely joined to a current light gathering to such an extent that the light radiating from the light can be observed. The light sensor itself is a hub in a heuristic remote interchanges organization. The light sensor capacities incorporate closest sensor neighbor location and recording, programmed correspondence course assurance to a hub regulator and giving continuous light status data to the hub regulator. The hub regulator itself is a light sensor with extra capacities, for example, correspondence capacity to a light Management Light Center, regularly by remote means. The hub regulator additionally contains the essential rationale to play out the control hub capacities.

The significance of traffic and walker security to specialized sufficiency assumes a significant part in each space. Set up road lighting plans use old strategies that need manual activity and need best in class advances. Accordingly, all labour's and power are problematic in the current worldview. For road lighting purposes, standard sodium fume lights are utilized. Regularly, these lights discharge a lot of warmth and gas into the climate. We utilize LED lights as an option of sodium fume lights in our proposed technique. Driven lights are right quite a bit longer than conventional lights. Both energy-proficient and practical LEDs. It additionally forestalls environmental emanations of ozone harming substances.

Driven bulbs utilize just 2-17watt of fluorescent energy. Besides, PWM (Pulse Width Modulation) framework to support computerize the light can screen the strength of the LED lights. In an area, traffic might be discontinuous, and streetlamp probably won't be fundamental for the whole width of the road. Re-appropriating the streetlamp will assist with saving a decent arrangement of power in these circumstances. We have in this way fostered a sharp road lighting gadget that contains a diming circuit graph that can fittingly raise or diminish the strength of streetlights from the outset, 20% of light sparkle at a lower power. At the point when the senor detects any vehicle, the light pressing factor rapidly ascends to its most significant level (i.e., 100%). We utilize the LDR to feel the power of the motor or in different terms the light delivered by its features is felt via vehicles.

GSM hardware screens the state of the thruway bridge. The gadget deformity is recognized by the sensor associated with every streetlight and sent a warning to the GSM module to invert the disappointment of the enlisted base station. This empowers the arranged plan to coordinate the new innovation to incorporate a shrewd streetlamp framework that diminishes upkeep expenses and expands the road lighting framework's presence.

In Section II, Device Design and Layout in Section III and in Sector IV application, discoveries and conversation in Section V, we tended to right off the bat comparable work Section VI Conclusion The climate has improved dynamically with the headway of innovation; metropolitan organization is basic to social, financial and actual wellbeing. Urban areas are in this manner confronting critical metropolitan improvement arranging difficulties. They face progressively expanding socioeconomics just as changes in friendly and dependability. Different strategies have been worked to robotize and work with shrewd urban communities.

Savvy urban communities are customarily addressed as complicated organizations made out of interdependencies between assets.

Late advancement in machine vision and keen innovation will assist with further developing local area recovery and decrease social advantages. We recommend a technique for recognizable proof of vehicle presence using a light associate system to improve day to day environments utilizing a set up PC and vision-based innovations. The gadget detects vehicles with streetlamps and recognizes empty leaving regions on the side of the road, cutting the degree of parking spot vital. It utilizes streetlamps to follow protests by consistently keeping a large portion of the light around evening time and changing to high iridescence.

Dependability is a primary thought to consider when a control methodology is chosen. Stable measurements can remember exact insights for road cycle paths, regardless of ecological factors like haze, brutal perspectives, and night climate. Furthermore, refreshing is significant, incorporating issues with the execution of current devices for everybody.

OBJECTIVES

- The target of the creation is to a development Intelligent Street and Parking Light Monitoring Through AI is a streetlamp clever observing and constant information control and the executives framework and approach for gathering sets of data from topographically appropriated areas worldwide, area, s or characterized sensor hubs committed to characterized streetlamps of a multi-light streetlamp framework and furthermore the innovation is to an energy utilization in metropolitan urban communities, little city, town and alternate way are developing step by step and in each city, a critical amount of power is being utilized for the possibility of road lighting another framework.
- The other target of the development is to It is a reality that the force use by streetlamps in certain pieces of the city is equivalent to that of regions with an extreme focus transducer, yet it is noticed that the volume of energy utilized by streetlamp in these areas is something very similar.
- The other goal of the innovation is to a Street lighting is a fundamental element of wellbeing plans in houses and business

establishments just as in outside parks, streetlamp the board has a vital capacity in a savvy observing plan that can be performed overwhelmingly in genuine checking conditions. In the PC share highlight different methodologies sent from the open organization PC from the Micro regulator.

- The other goal of the innovation is to the stargazing clock is broke down for experimental impacts. The unit has a LED light splendour controller, an ongoing clock, and a MOSFET circuit. While energy saving sounds wonderful, the plan actually has mishaps. The MOSFET has high impedance and is equipped for causing static force misfortune. Additionally, the obstruction of MOSFET is exceptionally low.
- The other goal of the innovation is to watch the ordinary power design for one year and adjust the iridescence of the street against PWM as per perceptions. This framework cap doesn't meet the unclear propensity or inconstancy in the climate.

SUMMARY

The development brings about colossal energy lost without use and LEDs will kill the proposed gadget, just as the high strength release lights, which can adjust its force by using LDR (Light Dependent Resistor) and which can change the direction of the vehicle by not utilizing LDR. The Device frequently faculties a framework disappointment and shows it on the base station using GSM innovations by submitting SMS (short message administration) Energy use is rising each day in metropolitan urban areas.

In each city, significant street light apparatus and a few segments of the city may be restricted recieving wires, yet the force utilized in these spots by lights is near that used by large radio wires. The current development is a gadget and system to screen traffic and traffic lights and to give ready notification to vehicles specifically. Fluffy is programming for consequently getting traffic signal parts from traffic streams and different factors, like climate changes; expected spikes in rush hour gridlock for top hours or extraordinary events.

This is a consequence of the time break in red and green for each traffic signal period. Ready signs are frequently shipped off engine vehicles as well as to traffic alerts, fixed or portable. The vehicles and additionally signs' GPS co-ordinates are not really set in stone through GPS satellite signs acquired. The alarm signs might include sporadic stage parts, subtleties on gridlocks, perilous situational data, fuel or synthetic breaks, gives an account of mishaps, and that's only the tip of the iceberg. Fluffy rationale controls measure risky alarm signs in the signs or vehicles and send pertinent notification to drivers based on the subtleties acquired and the real vehicle GPS or traffic notice image co-ordinates.

Fluffy rationale is then used to gauge traffic light parts and, contingent upon the deliberate stage parts and other traffic boundaries, compute OK danger alert signs. Fluffy conditions of thinking might be done at the traffic signals, at ready signs or in vehicles, either on a brought together force or on a scattered premise. Different center and scattered assessment variations may likewise be used. A profoundly coordinated, liquid based expert traffic the board gadget and interaction, which controls traffic lights and co-ordinates message control and signage, to additional improve traffic development and reduce blockage results.

The current innovation contains various units of live traffic gathering knowledge on traffic. Shrewd controls are accessible for the traffic knowledge units. Information from traffic is given to at least one focal controls. Holder boundaries and alarming subtleties will be dictated by the focal overseer or regulator. The fixation and ready information boundaries are shipped off the keen controls from at least one focal regulators. Brilliant controls are used to survey appropriate conduct based on pressure particulars and ready detail.

The current innovation identifies with a general framework and programmed interaction to distinguish and investigate the condition of each light in a lighting framework without adjustment of the current lighting framework electrical foundation. Explicit to the current executions of the above destinations that accommodate better current and opportune functional appraisal of constant capacities of said structure to address the issues of said goals of individual lights in a current lighting framework scattered over a geological region.

In one encapsulation, the innovation is a streetlamp observing and ongoing information the executives framework and procedure for gathering data from topographical conveyed areas or individual sensor hubs committed to individual streetlamps of a multi-light streetlamp framework involving: An astute identification sensor that performs self-checking to safeguard unwavering quality, accessibility and maintainability, can find close by or "neighbor" sensors, can find and set up a remote organization correspondence way, can independently change the remote organization correspondence way, and can take an interest in a correspondence convention for revealing estimated states of a given sensor.

A building mindful sensor equipped for fusing different kinds of sensors; an alludes to gathering information and to relegate little position as well as time arranges connected to the information gathered, such data for quick or relating transmission; a telecom for the exchange of information gathered; a sets up a connection b. A framework, alongside the information transmission, may get such data assembled from said assortment apparatuses and change such data accumulated into a logical work process and additionally other important information to change over such gathered information and relating space or potentially time co-arrangers.

The ascent of IoT has changed the way of life and upgraded different fields of society. The Artificial insight has expanded most urban areas across the world hope to become wise. The presentation of stopping the board arrangements, which permit residents to boost time, decrease fuel use and carbon dioxide emanations, is one of the most widely recognized use cases for savvy urban areas. Brilliant stopping frameworks with individual provisions have certain design.

Albeit just three viewpoints establish a shrewd stopping arrangement, it is important to take note of that every structure has a few distinct kinds of items which can be utilized for these methodologies. This archive diagrams the most widely recognized types, everything being equal, and represents designs being used all through the turn of events and foundation of study. It offers an extra perspective and an entirely significant information source. This information might be utilized by the clinical world to choose the assortment of the components to fuse an astute stopping arrangement. We look at here certain works applicable to the rollout of brilliant stopping advancements.

This postulation expects to extend the semi-quick strategy comparable to examining the point to accomplish this goal by doing a far reaching investigation just as an exploration on self-loader approach. The accompanying papers were inspected that discover normal subsections to help in adjusting the singular parts; the definitions are found in table structure to demonstrate how significant each paper is. each piece of the overview gets some information about the thing is presently occurring with sensors, boundaries, and programming items gets an update regarding what's coming later on The investigation diagrams extra elements to incorporate when fostering an insightful traffic arrangement outside client designs to incorporate a guide.

In the gadget adequately responds to the various controls sent by Zigbee module from a passage PC. It works in two primary modes, actual control and route framework. The bulbs are both flipped on or turned on during the intuitive control mode. The bulbs are stopped during daytime and turned on at evening in distantly controlled mode.

The gadget is exceptionally viable in saving power, yet it is fundamental to persistently overhaul the framework mode. With the utilization of a stroking module, it will be safer. A period subordinate power guideline has been set up to upgrade the energy upkeep of the streetlight. The stargazing clock is examined to give measurable outcomes in the particular geological district. The machine contains a microcontroller, a continuous clock and a MOSFET circuit to control the LED light power.

While the saving of energy sounds fabulous, there are additionally entanglements to the plan. The MOSFET has high impedance and low limit, making it helpless to static force harm. Likewise, MOSFET's resilience is extremely gentle. The frameworks track the customary daylight's force pattern more than one year and align the street light brilliance against PWM reliant upon perception results. As far as possible is this; it doesn't adjust to the unclear pattern or climatic variances.

DIAGRAM

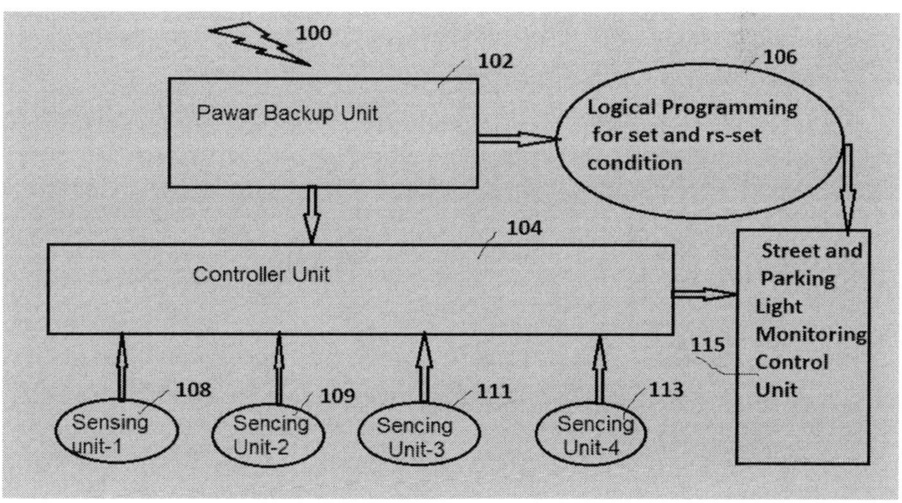

Fig. 1: Street light control block diagram.

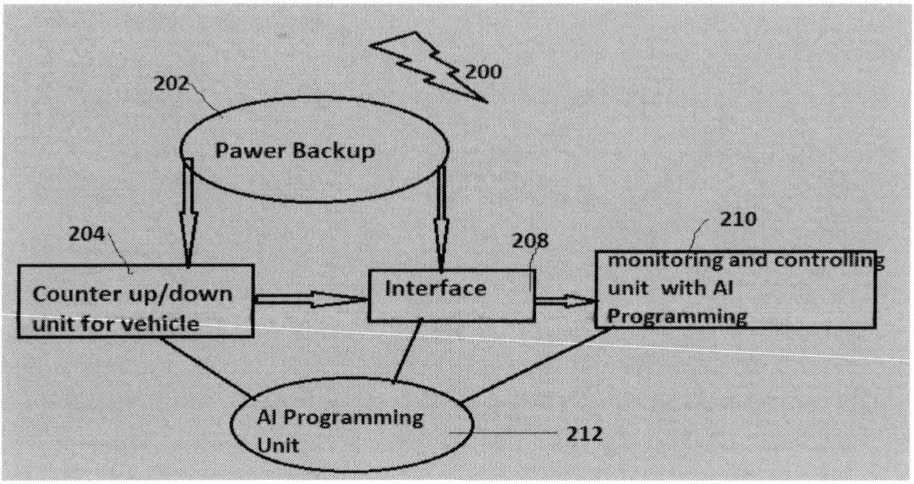

Fig. 2: parking light control block diagram

DESCRIPTION

A canny discovery sensor that performs self-checking to protect unwavering quality, accessibility and maintainability, can find close by or "neighbor" sensors, can find and set up a remote organization correspondence way, can independently change the remote organization correspondence way, and can take part in a correspondence convention for revealing estimated states of a given sensor; an insightful sensor engineering that obliges the capacity to fuse other sensor type;

In the most straightforward terms, information is crude material that is accumulated and a channel. Following that, information can be provided and followed by an organization association with terminals for assortment and terminals that permit the utilization of this data can be made accessible. Organizations can be made live through data terminals to quickly building up joins with others that would require it, or gave ahead of explicit time. Pre-recognized information organizing (out of the assortment interaction) in front of gathering results and supporting it to the pre-distinguished individuals with far off terminals to utilize the assembled data is fundamental.

A PC, alongside the information transmission, may get such information gathered from said assortment devices and change such gathered information into a logical work process or potentially other significant information to change over such information assembled and comparing space and additionally time co-organizes. The framework further involves implies for oversight the sharing of assets in light of the condition of the framework. The framework further includes a technique for acquire and dispensing data identifying with the detected circumstance.

Fig. 1: Street light control block graph.

Could you truly think like anything and furthermore how the streetlamps come on and go off all alone toward the day's end? Can somebody here/no one: Do any individuals on this island come on or off of the lights? There are numerous choices to enact the streetlamps, yet the accompanying circuit clarifies how an Automated Street Light Regulator actuates a LDR and a transfer for this work. For this situation, the gadget, a light/dim on/off switch is utilized, and may turn on or off streetlamps just as be applied to each home-grown contraption.

Fig. 2: Stopping light control block outline

This lighting and affectability (or all the more unequivocally, LDR) arrangement with the Parking Garage empowers the lights to be completely empowered when it is evening enough to utilize them. But since the ground pins of NPN are nonpartisan, the principal circuit schematic ought to be drawn for those included. When utilizing an undercarriage with a ground way of positive inside, the second cycle of the extended diagram will be drawn.

As the potentiometer (P1) is adjusted to its centre, the circuit begins to glow dimly. Printed circuit boards may be found in both the pleasant and unpleasant and negative halves of the chassis. Because the Transistors have various numbers per chassis, be aware of their types when connecting them to the chassis. These component numbers are described in the part location diagram as having similar wires:

The primary mechanical advantage of the engine live over the chassis is its strength any of the Park light bulbs. The connectors labelled in the parts placement diagram have the following connections:

P = Power live of the car

C = Car chassis

B = Park light bulb

OUTCOME

Our innovation Intelligent Street and Parking Light Monitoring Through AI is a streetlamp insightful checking and ongoing information control and the board framework and philosophy for gathering sets of data from topographically circulated areas worldwide, area, s or characterized sensor hubs devoted to characterized streetlamps of a multi-light streetlamp framework. The development is energy utilization in metropolitan urban communities, little city, town and alternate way are developing step by step and in each city, a huge amount of power is being utilized for the possibility of road lighting another framework. It's obviously true that the force use by streetlamps in certain pieces of the city is equivalent to that of regions with an extreme focus transducer, however it is noticed that the volume of energy utilized by streetlamp in these areas is something similar. The development is in like manner a Street lighting is a fundamental component of security plans in houses and business establishments just as in outside parks, streetlamp the board has a critical capacity in a shrewd checking plan that can be performed overwhelmingly in genuine observing conditions. In the PC share highlight different methodologies sent from the open organization PC from the Micro regulator. It works in two essential modes, the strategy for actual guideline and directing. Through cooperative control mode, the lights are both turned on or on. As the day progressed, the lights are turned off in physically worked mode, and are empowered at dim. The organization is extremely fruitful in power preservation, yet it is fundamental that the framework mode is persistently improved. It would be more secure with the utilization of a strobing module. To improve the energy protection of the streetlight, a worldly power control has been made. In a similar topographical region, the cosmology clock is dissected for exact impacts. The unit has a LED light brilliance controller, a continuous clock, and a MOSFET circuit. While energy saving sounds amazing, the plan actually has difficulties. The MOSFET has high impedance and is equipped for causing static force misfortune. In addition, the opposition of MOSFET is extremely low. The frameworks watch the typical force design for one year and align the iridescence of the street against PWM as indicated by perceptions. This framework cap doesn't meet the unclear inclination or changeability in the climate.

1. According to claim1# the innovation is to a creation Intelligent Street and Parking Light Monitoring Through AI is a streetlamp savvy checking and continuous information control and the board framework and technique for gathering sets of data from topographically appropriated areas worldwide, area, s or characterized sensor hubs devoted to characterized streetlamps of a multi-light streetlamp framework and furthermore the development is to an energy utilization in metropolitan urban communities, little city, town and alternate way are developing

step by step and in each city, a huge amount of power is being utilized for the possibility of road lighting another framework. It's undeniably true that the force use by streetlamps in certain pieces of the city is equivalent to that of regions with an extreme focus transducer, however it is noticed that the volume of energy utilized by streetlamp in these areas is something very similar.

2. According to claim1,2,3# the innovation is to a Street lighting is a fundamental component of wellbeing plans in houses and business establishments just as in outside parks, streetlamp the board has a vital capacity in a shrewd checking plan that can be performed energetically in genuine observing conditions. In the PC share highlight different methodologies sent from the open organization PC from the Micro regulator.
3. According to claim1,2,3# the creation is to the stargazing clock is dissected for observational impacts. The unit has a LED light splendor controller, an ongoing clock, and a MOSFET circuit. While energy saving sounds marvellous, the plan actually has mishaps. The MOSFET has high impedance and is fit for causing static force misfortune. Besides, the obstruction of MOSFET is extremely low. The frameworks watch the ordinary power design for one year and align the iridescence of the street against PWM as indicated by perceptions. This framework cap doesn't meet the vague inclination or changeability in the environment.

REFERENCES

[1] Abhigna, B.S., Chakraborty, S., Kumar, V., Medha: A collaborative E-concept framework, Proceedings of the 2018 International Conference On Communication, Computing and Internet of Things, IC3IoT 2018, 2019, pp. 117–122, 8668127

[2] Abidin, S., Vadi, V.K., Rana, A., On Confidentiality, Integrity, Authenticity, and Freshness (CIAF) in WSN, Advances in Computer, Communication and Computational Sciences, 2020, pp 87-97

[3] Ahay, A., Choudhury, T., Kumar, V., Hierarchical Evolution of Digital Games, Proceedings of 2019 International Conference on Computational Intelligence and Knowledge Economy, ICCIKE 2019, 2019, pp. 381–386, 9004378

[4] Area: Communication, Frequency modulation, Networks, wireless communication

[5] Area: Communication, Frequency modulation, Networks, wireless communication

[6] Area: Communication, Frequency modulation, Networks, wireless communication

[7] Area: Communication, Frequency modulation, Networks, wireless communication

[8] Arora, M., Kumar, V., Dave, M., Hybrid cloud infrastructure for load balancing and scaling with vm path optimization using ant colony optimization, Journal of Advanced Research in Dynamical and Control Systems, 2020, 12(5 Special Issue), pp. 660–674

[9] Arora, M., Kumar, V., Dave, M., Task scheduling in cloud infrastructure using optimization technique genetic algorithm, Proceedings of the World Conference on Smart Trends in Systems, Security and Sustainability, WS4 2020, 2020, pp. 788–793, 9210303

[10] Bansal, R., Gupta, A., Singh, R., Nassa, V.K., Role and impact of digital technologies in E-learning amidst COVID-19 pandemic, Proceedings - 2021 4th International Conference on Computational Intelligence and Communication Technologies, CCICT 2021, 2021, pp. 194–202

[11] Bansal, R., Obaid, A.J., Gupta, A., Singh, R., Pramanik, S., Impact of Big Data on Digital Transformation in 5G Era, Journal of Physics: Conference Seriesthis link is disabled, 2021, 1963(1), 012170

[12] Chauhan, S.S., Khare, A.K., Kinematic analysis of the ABB IRB 1520 industrial robot using roboanalyzer software, Evergreenthis link is disabled, 2020, 7(4), pp. 510–518

[13] Chauhan, S.S., Khare, A.K., Method of Obtaining Planar State of Stress Using Mohr's Circle—Some Typical Cases, Lecture Notes in Mechanical Engineering, 2022, pp. 33–38

[14] Goel, S., Tripathi, A., Choudhury, T., Kumar, V. Parkinson's Disease Detection using Soft Computing Technique, Proceedings of the 2019 8th International Conference on System Modeling and Advancement in Research Trends, SMART 2019, 2020, pp. 3–6, 9117384

[15] Govindarajan, V., Sivakumar, R., Patil, P.P., Kannan, M., Ramesh, B., Effect of Tungsten Carbide Addition on the Microstructure and Mechanical Behavior of Titanium Matrix Developed by Powder Metallurgy Route, Advances in Materials Science and Engineeringthis link is disabled, 2022, 2022, 2266951

[16] Gupta, A., Singh, R., Nassa, V.K., Sharma, P., Koti, K., Investigating Application and Challenges of Big Data Analytics with Clustering, 2021 International Conference on Advancements in Electrical, Electronics, Communication, Computing and Automation, ICAECA 2021, 2021

[17] Jain, P., Vijay, S., & Gupta, S. C. (2009). ATM Network QoS Parameters and Performance Analysis using OPNET Simulator. International Journal of Recent Trends in Engineering, 2(4), 228.

[18] Jain, P., Vijay, S., & Gupta, S. C. Performance Analysis & QoS Guarantee in ATM Networks. Global Journal of computer Science and Technology, 131-136.

[19] Jain, S., Choudhury, T., Kumar, V., Kumar, P., Detecting malware and analysing using sandbox evasion, Proceedings of the 2018 International Conference On Communication, Computing and Internet of Things, IC3IoT 2018, 2019, pp. 111–116, 8668149

[20] Kala, S., Nandan, H., Sharma, P., Shadow and weak gravitational lensing of a rotating regular black hole in a non-minimally coupled Einstein-Yang-Mills theory in the presence of plasma, European Physical Journal Plusthis link is disabled, 2022, 137(4), 457

[21] Kandpal, H., Kumar, A., Malik, A., Morphometric analysis and prioritization of sub-watersheds in Bino watershed, Uttarakhand: A remote sensing and GIS perspective, Indian Journal of Ecology, 2018, 45(3), pp. 475–478

[22] Kannan, M., Singh, S., Prasad, R.R., Synthetic Methods to Obtain Calcia-Stabilized Zirconia Powders—A Review, Lecture Notes in Mechanical Engineeringthis link is disabled, 2021, pp. 405–416

[23] Kukreti, A., Painolli, A.K., Rana, N., Where do we stand: factors affecting sustainable development, Procedia Environmental Science, Engineering and Management, Volume 8, Issue 2, 2021, Pages 583-589

[24] Kulkarni, S & Chima, P 2021 'Challenges faced by UK university students due to thcoronavirus crisis in the Higher Education' Preprints.org. https://doi.org/10.20944/preprints202102.0192.v1

[25] Kulkarni, S & Sodre, JR 2017, An Assessment of Lightweight Vehicles to Reduce Greenhouse Gas Emissions with Focus on Magnesium. in Thoughts and Refections on the use of Magnesium. Birmingham City University, pp. 84-91.

[26] Kulkarni, S, Chapman, C & Shah, H 2016 'Computational Fluid Dynamics (CFD) Mesh Independency Study of A Straight Blade Horizontal Axis Tidal Turbine' Preprints.org. https://doi.org/10.20944/preprints201608.0008.v1

[27] Kulkarni, S, Chapman, C, Shah, H & Edwards, D 2018, 'A computational design method for bio-mimicked horizontal axis tidal turbines', International Journal of Building Pathology and Adaptation, vol. 36, no. 2, pp. 188-209. https://doi.org/10.1108/IJBPA-06-2017-0029

[28] Kulkarni, S, Chapman, C, Shah, H, Parn, E & Edwards, DJ 2018, 'Designing an efficient tidal turbine blade through bio-mimicry: A systematic review', Journal of Engineering, Design and Technology, vol. 16, no. 1, pp. 101-124. https://doi.org/10.1108/JEDT-08-2017-0077

[29] Kulkarni, S, Chapman, C, Shah, H, Parn, E & Edwards, DJ 2018, 'Design Study of Horizontal Axis Tidal Turbine Blade', Mindanao Journal of Science and Technology. https://www.open-access.bcu.ac.uk/10089/

[30] Kulkarni, S, Edwards, DJ, Chapman, C, Hosseini, MR & Owusu-Manu, D-G 2019, 'A preliminary mechanical design evaluation of the Wikispeed car: for light-weighting implications', Journal of Engineering, Design and Technology, vol. 17, no. 1, pp. 230-249. https://doi.org/10.1108/JEDT-09-2018-0154

[31] Kulkarni, S, Edwards, DJ, Parn, E, Chapman, C, Aigbavboa, CO & Cornish, R 2018, 'Evaluation Of Vehicle Lightweighting To Reduce Greenhouse Gas Emissions With Focus On Magnesium Substitution', Journal of Engineering, Design and Technology, vol. 16, no. 6, pp. 869-888. https://doi.org/10.1108/JEDT-03-2018-0042

[32] Kulkarni, SS, Wang, L, Golsby, N & Lander, M 2021, 'Fluid-structure interaction based optimisation in tidal turbines: A perspective review',

Journal of Ocean Engineering and Science. https://doi.org/10.1016/j.joes.2021.09.017

[33] Malik, A., Kumar, A., Kandpal, H., Morphometric analysis and prioritization of sub-watersheds in a hilly watershed using weighted sum approach, Arabian Journal of Geosciencesthis link is disabled, 2019, 12(4), 118

[34] Mohanakurup, V.K., Baghela, V.S., Kumar, S., Srivastava, P.K., Doohan, N.V., Soni, M., Awa, H., 5G Cognitive Radio Networks Using Reliable Hybrid Deep Learning Based on Spectrum Sensing, Green Internet of Things in 5G and Beyond, Hindawi, Volume 2022

[35] Mohta, M., Soni, N., Choudhury, T., Kumar, V., Market Analysis on Li-Fi Technology, Proceedings of the 2019 8th International Conference on System Modeling and Advancement in Research Trends, SMART 2019, 2020, pp. 139–143, 9117427

[36] Nagwan, H., Rana, N., Kumar, A., Tewari, P.C., Performance and emission characteristics of diesel engine using Karanja oil methyl ester and Eucalyptus oil fuel blends, Procedia Environmental Science, Engineering and Management, Volume 6, Issue 4, 2019, Pages 553-563

[37] Painoli, A.K., Bansal, R., Singh, R., Kukreti, A., Impact of digital marketing on the buying behavior of youth with special reference to Uttarakhand state, Big Data Analytics for Improved Accuracy, Efficiency, and Decision Making in Digital Marketing, 2021, pp. 162–182

[38] Parag, J. (2011). Hybrid mixed handover for call blocking probability analysis in wireless ATM network. Journal of Engineering and Computer Innovations, 2(4), 68-77.

[39] Sharma, C., Singh, M.P., Chaudhary, S., Cost Analysis of Solar-Wind Based Hybrid Renewable Energy System, Lecture Notes in Electrical Engineeringthis link is disabled, 2021, 700, pp. 2733–2746

[40] Sharma, S., Chandra, U., & Jain, P. (2017). A literature survey on automation of test data generation for branch coverage testing using genetic algorithm. International Journal of Computational Intelligence Research, 13(6), 1521-1531.

[41] Varshney, R., Jain, P., & Vijay, S. (2018, September). Massive MIMO Systems In Wireless Communication. In 2018 2nd International Conference on Micro-Electronics and Telecommunication Engineering (ICMETE) (pp. 39-44). IEEE.

CHAPTER - 14

AI - Based Intelligent Robotic (Robot Bird Flying System)

Dr. Sanjay B. Waykar
Mahatma Gandhi Mission's College of Engineering and Technology, India. Email:sbwaykar@gmail.com

Dr. Anil W. Kale
Mahatma Gandhi Mission's College of Engineering and Technology, India. Email: anil5474@gmail.com

Dr. Nandkishor P Karlekar
MGMCET Kamothe Navimumbai, India. Email: npkarlekar.mgmcet@gmail.com

Dr. Vaibhav Narawade
Ramrao Adik Institute of Technology, India. Email:vnarawade@gmail.com

ABSTRACT

Our Invention "Clever ROBOTIC (Robot Bird Flying System) "is a robot bird flying controllability assessment framework and assessment technique and the assessment framework, a microchip B produces boost beat arrangements with continuously changed strength as indicated by the preset guideline, a chip and is responsible for GPS information examination and applicable data record, the chip B helps out the chip A. The creation is the pertinent test information assortment of a robot bird to be identified are finished in a coordinated way, the gathered information is prepared through an information examination arrangement of a PC prior to being changed over into a direction dependent on an electronic guide, if the robot bird has the flying controllability can have decided based on the direction and the significant data, and the assessment if the robot bird is qualified can be finished. The development is to a thorough assessment of the robot bird flying controllability can be understood, the earlier data needed by flight route can be gained, the visual deficiency of resulting flight route tests is decreased, a lot of invalid redundant work is stayed away from, the functioning proficiency is improved incredibly, and the robot bird useful advancement can be sped up. The innovation additionally incorporates a technique for a robot bird, which takes on an appropriate miniature incitement sign to opportune animate a stomach dorsal middle core and a robustus archistriatalis of a thalamencephalon of the bird to lead a functioning getting away from conduct of the bird; in this way directing the flying conduct of the bird. The development uses the dynamic getting away from conduct of the bird came about because of utilizing a miniature power to reproduce a particular nerve cores of the bird, specifically utilizing a rule of utilizing the miniature power to recreate the nerve cores to picture 'hurt' to direct the conduct of the bird. The strategy utilized for directing the robot bird of the creation enjoys the benefits that as the standard uses the appetence of the bird to keep away from hurt, the innovation has an incredible directing capacity; along these lines requiring not to prepare the bird ahead of time; the robot bird has a gigantic utilizing esteem. The creation is a contrasted and the ground creature, the bird has a bigger dynamic space and quick speed, in this manner having a bigger application advantage clearly the reasonable and potential application is exceptionally wide and an advanced fly motor bay assurance framework that ensures against enormous birds and works independently, that is torpid in routine airplane activities. The development take a naturally activates its defensive device(s) just at the prompt mark of need, then, at that point returning it (them) to a non-meddling position, including a RADAR framework and a LIDAR framework which recognize birds entering the expected flight way of the airplane, and quick PC executed computational calculations that track and distinguish those from the identified set that are (1) projected to enter a zone which would prompt ingestion by the motor, and (2) of a size enormous enough that they would genuinely harm the motor whenever ingested. The development likewise incorporates guarded systems housed in the motor nacelle cowling or focus center, or in the fuselage structure that are promptly impelled in a matter of seconds prior to showing up at the focal point to safeguard, avoid, decrease the size of the moving toward bird to a satisfactory mass, or annihilate it, and, that after the ingestion danger has passed, the gadgets are stowed/withdrawn or securely casted off.

FIELD

The creation "Clever ROBOTIC (Robot Bird Flying System) " is identified with a savvy robot bird flying controllability assessment framework and furthermore identifies with frameworks and strategies for shielding plane motors from ingestion of unfamiliar items, like huge birds, during flight and an elective exemplification explicitly adjusted for assurance of a weak locale of helicopters. All the more especially the frameworks identify with mechanical gadgets that capacity to relieve the danger of bird strikes to airplane motors and the creation identifies with creature robot control field, particularly a sort of robot bird in free space, fly assessment framework and the appraisal strategy for controllability.

BACKGROUND

Creature robot is to accept living creature as body, control its cerebral nerve or muscle with coded electrical sign, to understand the savvy creatures that body is controlled. Robot bird is a sort of in creature robot, for instance, accept pigeon as body, can handle the engine conduct of pigeon, in this way cause pigeon to turn into a controlled machines individuals bird by the nerve core that miniature electrical incitement is applicable. Its standard of work is as per the following:

In robot bird configuration measure, will create a ton of dubious components: the blunder in surgery, the discreteness of the distinction of creature individual self and the production art of animating cathode. Previously mentioned joined variables impact will cause for various robot birds, their controllability, and affectability and required boost power are unique; For same robot bird, the necessary appropriate improvement force of various nerve core is likewise discrepant. Indeed, can cause created robot bird not have controllability, be an unsatisfactory item. Therefore, before flight route analyze, to a robot bird, it is exceptionally vital that the robot bird of especially recently creating does test and evaluation completely.

Additionally, don't have at connection procedure and technique to do quick exhaustively flight controllability test and evaluation to robot bird. Need simply test and examination innovation and technique, just can't acquire the earlier data of tried robot bird. Cause coming up short on the far reaching evaluation to tried robot bird controllability, do not have the appraisal of the comparing connection of controllable affectability and boost power, likewise do not have the characterizing of reasonable upgrade force extent of three nerve core of tried robot bird. This causes the flight route analyze device of robot bird to bear the personality of much visual deficiency, and experimentation created a lot of invalid rehashed works, and strategy is science and shortcoming not.

Different plans proposed utilizing a screen having versatile individuals, permitting the screen to be in a kind of withdrew mode for what was viewed as non-unsafe segments of the flight. Notwithstanding, such plans introduced a few issues, including exceptionally complex systems required for development between their put away and functional states. Additionally, despite the fact that birds are bound to be experienced via airplane during the lower elevation flight leaving or showing up at an air terminal, there have been many harming bird strikes in other flight systems, and those plans don't give airplane motor insurance all through the whole flight envelope.

OBJECTIVES

- The target of the innovation is to an is robot bird flying controllability assessment framework and assessment technique and the assessment framework, a microchip B produces boost beat arrangements with steadily changed strength as per the preset standard, a chip An is responsible for GPS information examination and pertinent data record, the microchip B helps out the microchip An and the development is the applicable test information assortment of a robot bird to be distinguished are finished in a collaborated way, the gathered information are handled through an information investigation arrangement of a PC prior to being changed over into a direction dependent on an electronic guide, if the robot bird has the flying controllability can decided based on the direction and the significant data, and the assessment if the robot bird is qualified can be finished.
- The other target of the innovation is to an exhaustive assessment of the robot bird flying controllability can be understood, the earlier data needed by flight route can be obtained, the visual impairment of ensuing flight route tests is diminished, a lot of invalid monotonous work is kept away from, the functioning effectiveness is improved significantly, and the robot bird down to earth progress can be sped up.
- The other target of the innovation is to the robot bird has a colossal utilizing esteem. The innovation is a contrasted and the ground creature, the bird has a bigger dynamic space and quick speed, in this manner having a bigger application advantage clearly; the sensible and potential application is extremely wide and an advanced stream motor channel security framework that ensures against enormous birds and works independently, that is lethargic in routine airplane tasks.
- The other target of the innovation is to a consequently activates its defensive device(s) just at the prompt mark of need, then, at that point returning it (them) to a non-meddling position, including a RADAR framework and a LIDAR framework which

recognize birds entering the planned flight way of the airplane, and quick PC carried out computational calculations that track and distinguish those from the identified set that are (1) projected to enter a zone which would prompt ingestion by the motor, and (2) of a size huge enough that they would genuinely harm the motor whenever ingested.

- The other target of the development is to a protective systems housed in the motor nacelle cowling or focus center, or in the fuselage structure that are promptly activated presently prior to showing up at the focal point to safeguard, redirect, diminish the size of the moving toward bird to an OK mass, or obliterate it, and, that after the ingestion danger has passed, the gadgets are stowed/withdrawn or securely discarded.

SUMMARY

The motor security framework, for which different encapsulations are depicted thus, in its put away mode doesn't debase airplane execution, conveys just when required and is impelled in a convenient manner because of data it assembles about approaching dangers, principally birds. They give defensive gatekeepers to stream airplane motors, and further developed techniques for recognizing, distinguishing, following, and foreseeing the probability of crash with a secured motor by bird(s) of still up in the air to require ensuring against sway; deciding the need of conveying the guard(s) against such bird(s) instead of moving the plane to keep away from the bird totally; impelling and sending the securing device(s), in case not really settled to do as such, in adequate opportunity to turn away or redirect sway with the motor; and for withdrawing or securely arranging the device(s); all including, however not restricted to, the advanced high-sidestep proportion turbofan fly motor, and to ensuring the uncovered, weak parts of a helicopter rotor control framework against the particular dangers recorded previously.

In substitute encapsulations they accommodate consequently securing the air admission of a cutting edge high-sidestep proportion turbofan motor from the ingestion of birds and other unfamiliar items. The favored framework is ideally housed in the nacelle cowling or focus center point of the turbine motor or fuselage structure which, on activation, grows to immediately safeguard, divert, lessen the size of the moving toward bird or other unfamiliar items to an OK mass. Incitation is ideally instructed consequently, without requiring human choice or activity, by a sub-framework involved a sensor which recognizes birds or other article entering the expected flight way of the airplane; quick computational calculations that track and distinguish those from the identified set that are projected to enter a zone which would prompt ingestion by the motor, and of a size or weight adequately huge to truly harm the motor whenever ingested.

After the ingestion danger has been discarded, parts of the framework are stowed, withdrawn or securely casted off. The current frameworks give elective encapsulations to managing enormous birds and furthermore securing the center motor locale of the delta. The current frameworks hence keep certain unfamiliar items from entering the air admission of a stream turbine motor. At long last, the frameworks additionally incorporate an elective exemplification that is explicitly helpful for and adjusted to insurance of a weak district on helicopters.

The frameworks identify with defensive gatekeepers for stream airplane motors, explicitly to further developed strategies for recognizing, distinguishing, following, and anticipating the probability of crash with a secured motor by bird(s) of not really settled to require ensuring against sway; deciding the need of conveying the guard(s) against such bird(s); inciting the ensuring device(s) in adequate chance to turn away or divert sway with the motor; and for its withdrawal or safe removal; all including, however not restricted to, the cutting edge high-sidestep proportion turbofan fly motor, and to insurance of the uncovered, weak parts of a helicopter rotor control framework, against the particular dangers recorded previously.

After this, microchip A gets and protects gps information to TF card stockpiling unit, in the meantime, separates the speed data involving in gps information outline; If speed is more prominent than 1 meter each second, represent that robot bird is in condition of flight, can fly and control test tests; Now first appointed authority whether this test finishes, if complete, microcontroller A drops flight and controls test sign; If don't finish, chip An organizes flight control test sign notification microchip B can execute flight control test.

In case speed isn't more than 1 meter each second, outline that robot bird, not in condition of flight, presently can drive the test tries that takes off; Equally, first whether judgment is compelled to take off to test and is finished, if complete, microcontroller A drops powers to take off test sign; If don't finish, chip A masterminds the test sign notification microchip B that powers to take off can execute to power to take off test; When flight, control test and pressing factor and take off after test all finishes, microchip A completions all working.

The appraisal procedure of robot bird flight controllability is as follows:

By the investigation to direction, if the bearing of movement of robot bird isn't liable to acquire the effect that animates, check boost power data relating to segment direction therewith, if upgrade force has arrived at maximal worth, outline that this robot bird doesn't have flight controllability.

By the examination to direction, in case be a bend turning left relating to invigorating the direction of left DIVA, and can rehash to get the above outcomes, the left DIVA that robot bird is depicted is that flight is controlled. In like way, be a bend turning right, and can rehash to get the above outcomes comparing to animating the direction of right DIVA, the right DIVA that robot bird is depicted is that flight is controlled.

Investigation dependent on to previously mentioned controlled direction, can get the controllable affectability of tried robot bird and the comparing connection of upgrade force, and decides the viable boost strength scope of relating nerve core.

The assessment arrangement of robot bird given by the creation and appraisal strategy, assessment framework finishes assortment and the examination of information, and various and different information are changed over into designed outcome; Based on this graphical outcome and related to evaluation technique for the current development, can finish the evaluation whether tried robot bird has flight controllability. This apriority data will precisely educate the subsequent flight route test of this robot bird, along these lines the visual deficiency of keeping away from and rehashed non-useful work are expanded work proficiency enormously.

The target of the innovation is to defeated the lack in existing robot creature route innovation and the technique, give a sort of and need not the bird precondition, the direction work is solid, and the utilization esteem stature helps the strategy for direction that the robot creature enters the robot bird of viable stage. For arriving at reason, a sort of technique for direction that is utilized for robot bird given by the innovation, this strategy is executed control by containing the direction framework that is made of PC controlled cell site and miniature incitement signal generator to the flare move and the course of robot bird, and this strategy is completed after advances:

1. at portrayed robot bird back or crown segment place the miniature incitement signal generator;

2. by depicted by PC controlled cell site, the miniature incitement signal generator

The benefit that the current development is utilized for the strategy for direction of robot bird is: attributable to used previously mentioned nerve core by miniature electrical incitement empty head to cause its initiatively get away from conduct its engine conduct of controlling and explore, expeditiously used bird to get away from the intuition of injury, so its direction work is solid, consequently need not bird is done precondition. The enormous use esteem that robot bird has, birds are contrasted and ground creatures, and greater action space, development speed are quicker masterminded, and greater application advantage is clearly orchestrated, and the truth and possibility reasons for existing is generally. For instance, the bird that has a specific microchannel sensor can be utilized for the assortment of view, presentation and natural boundary and so on in explicit (or target) zone and pass back and so forth; Can be utilized for uncommon article shipped off or be tossed into objective or something like that. When flying distance surpasses " locating distance ", then, at that point can be by its flight way of GPS route. Since flight is simply the actual productivity by birds, so its flight time and flying distance simply are exposed to the limitation of independent energy capacity dissimilar to the microminiature automated plane. Clearly, the portable robot contrasts and electromechanical, and the robot bird that acquires as indicated by the current development will enjoy gigantic benefit.

DIAGRAM

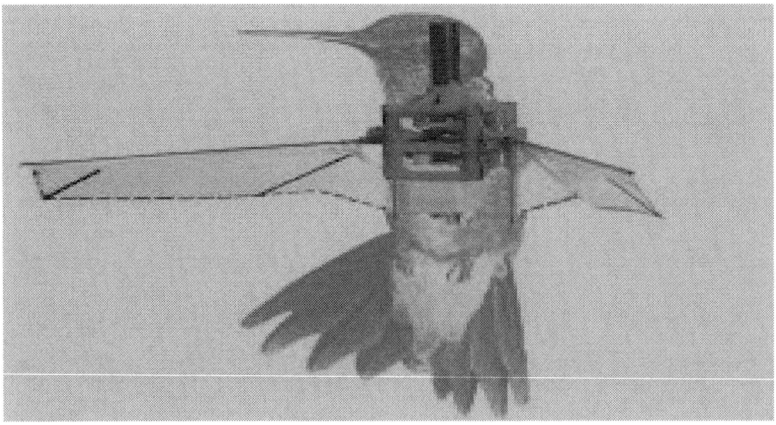

Fig. 1: is a simplified schematic depicting the distributed components of several embodiments of a preferred jet engine protection system

Fig. 2: is a front view of a modern, conventional high-bypass ratio turbofan jet engine;

Fig. 3: is a simplified cross-sectional view of a modern, conventional high-bypass ratio turbofan jet engine;

DESCRIPTION

The light of this assessment framework and the made determination of evaluation technique, explores can get the earlier data of tried robot bird. As per these data, robot bird probably can be isolated into three kinds: confirmed items, unsuitable item and imperfection product. Explores figures research approach focused on as indicated by the kind under robot bird, consequently, stay away from visual deficiency, increment work productivity. Substantial plan is as per the following:

(1) whenever tried robot bird is confirmed items, direction mirrors that three nerve core of tried robot bird all have great controllability. The flex point situated in direction, the side of examination robot bird is additionally checked comparing improvement power data, can acquire an applicable boost force comparing to tried robot bird and the standard of corner, it has mirrored the connection of controlling affectability and upgrade force; Meanwhile, can decide the appropriate upgrade force extent of each nerve core. These are the significant earlier data's before tried robot bird flight route, and earlier data can be stayed away from the visual deficiency of flight route analyze, enormously builds work proficiency.

(2) whenever tried robot bird is inadmissible item, direction doesn't mirror that tried robot bird has any controllability. For this robot bird, don't have to do again flight route try, yet examine focused on and search reason, aggregate insight, summarize illustration, accordingly stay away from invalid rehashed work.

As a rule, the sensor part or sub-framework identifies an approaching item in the airplane's flight way and afterward conveys information messages to the evaluator, which regularly is an advanced PC or computerized processor, most ideally a continuous inserted programming program. The evaluator decides the probability of the article's crash with the airplane and afterward produces yield control signs to activate the mechanical obstruction sub-framework or gadget. These frameworks can include and utilize, notwithstanding all around created materials innovations, present day traditional advances, for example, bird recognition radar, scaled down PCs using quick computational calculations, parts, for example, activating instruments and materials from car accident airbag frameworks, progressed receptive cautious framework parts, and, alternatively, microwave innovation or other coordinated energy techniques.

The frameworks depicted in this are expected to ensure an advanced fly motor against generally enormous birds or different items which, whenever ingested would truly harm or obliterate the motor. The dynamic parts or individuals from these frameworks are torpid during routine airplane activities, and ideally are consequently impelled just when a quick need, i.e., inevitable crash emerges. After use in flight, the dynamic individuals are then gotten back to their lethargic, non-meddling position or arranged.

As a general rule, the general framework ideally incorporates a sensor sub-framework which recognizes birds entering the expected flight way of the airplane; PC executed quick computational calculations that track and distinguish such birds or items from the identified set that are projected to enter a zone which would prompt ingestion by the motor and are of a size sufficiently huge or weight extraordinary enough that they would genuinely harm the motor whenever ingested, and a transitory cautious framework, like a mechanical watchman gadget, a move schedule, or a coordinated energy producer. In a matter of seconds before the recognized impediment is to show up at the focal point, the flitting cautious framework is turned on and acts immediately to safeguard, avoid, decrease the estimate or obliterate the moving toward bird so it or its remaining parts are diminished to mass that won't make genuine harm the motor. After the ingestion danger has passed, the dynamic gadgets are ideally stowed, withdrew, securely discarded, or deactivated.

Organization Requirements

To comprehend the heaps coming about because of the effect of a bird, the term of the impact occasion, now and again alluded to as the "squash-up time", still up in the air. Test results show that this occasion generally endures 3-6 milliseconds. In view of the delegate airplane and bird type, the circumstance prerequisites for sending of the framework not set in stone. Assume a 747-class airplane is cruising at 555 miles each hour (48,840 feet each moment) and experiences a Canadian goose in its flight way. In the event that the bird is flying at 3,500 feet each moment and the sensor sub-framework can distinguish its quality at a scope of 0.5 miles, then, at that point the time among recognition and impact would be roughly 3 seconds. In the event that a security factor is remembered for the computation, the arrangement time for the moderation gadget would be close to 1-2 seconds. To meet this organization necessity, damped pyro actuators would be appropriate, as would be a direct actuator.

Sensor Sub-Systems

Ordinary RADAR and LIDAR frameworks have the ability for playing out the sensor capacities in the current frameworks and are hence viewed as favored parts in a sensor sub-framework. As utilized in this setting the term LIDAR is planned to have its usually got meaning: a gadget that is comparative in activity to radar however radiates beat laser light rather than microwaves. A LIDAR situation like the Rigel VQ-480 situation has a shaft impression of around 9 creeps in measurement a way off of one-half mile (viewed as 2,600 feet). To enlist an article on this framework, the objective should be bigger than the impression of the bar or continuous sweeps should be made to cover and hence to make a successfully more modest impression.

For a bird with a front facing outline of 12 crawls in measurement, a moderate gauge for a Canadian goose, the LIDAR shaft will effortlessly enroll the objective at this distance without expecting to cover filters. In the event that the pillar is cleared across its 60-degree field of view, that is 90 feet above and beneath the airplane, which is 2 degrees above and underneath the projected flight way, at an output pace of 100 sweeps each second, it will require almost 2.5 seconds to clear across the whole space of interest. Since this output time is on a similar request as the time accessible for the ostensible case arrangement time, it is accepted that a LIDAR based sub-framework will work effectively as a sensor part or sub-framework. LIDAR items with longer identification ranges are as of now economically accessible, and this innovation is probably going to progress in the coming years, further upgrading the affectability of the framework.

Mechanical Deflection Sub-Systems

The initially favored mechanical encapsulation or sub-framework incorporates a bunch of smoothed out poles or cylinders separated equitably around the periphery of the nacelle lip at its forward or driving end. They are ideally housed in the motor nacelle cowling, focus center, or airplane fuselage structure, and can be impelled to in a split second send straightforwardly before the motor. At the point when deactivated, this encapsulation can be stowed by getting back to its unique arrangement situated inside the motor nacelle cowling, focus center, or airplane fuselage structure.

These poles are expected to capacity to redirect the approaching avian danger away from the motor or to section the corpse to a size that doesn't compromise the motor. The poles or cylinders are ideally smoothed out and equitably dispersed around the perimeter of the nacelle. At the point when conveyed they meet at or close to a solitary point before the motor. The bars are stowed in a withdrawn situation toward the front of the motor nacelle, and are impelled forward when a bird sway is fast approaching. When the bird danger has died down, the bars are withdrawn once again into the nacelle to allow unhindered wind current to enter the motor.

The poles are ideally sent in both of two different ways: first, by a mechanical activation gadget driven by an engine; or second, by a pyrotechnically impelled gadget. It is imagined that the pyrotechnical gadget will require less mass and will be simpler to incorporate into the airplane nacelles. To send straight forward toward the front of the motor, a damped pyrotechnic actuator would touch off to push the poles through a bunch of guide openings. The aide openings would capacity to guarantee arrangement. The framework would likewise be built so the aide openings would constrain the main finishes of the poles to meet or verge on gathering at a typical vertex. The sending time would be on the request for one second.

Nitty gritty investigation of bird effects can be found in the scholarly writing. In many investigations, the bird has been displayed as a delicate body and treated as either a uniform liquid, like water, or as a mass with variable thickness, like water with air bubbles. To appraise the powers which the cone of bars or cylinders may insight during a bird sway, one can consider a bird slamming into a solitary cylinder, and at the cylinder's midpoint in order to introduce the direst outcome imaginable. At the point when the bird slams into the pole, the bar encounters a close prompt shock pressure followed by a stagnation pressure. The two occasions together last an aggregate of a couple of milliseconds.

In one favored epitome the cone of bars or cylinders is increased with extra primary individuals close to the nacelle lip. These individuals alleviate the chance of an ingestion around here, where the bar dividing may surpass the components of the bird's body, hence leaving an opening large enough for the bird to fall through. These extra individuals would at first send with the parent bars and afterward be collapsed outward utilizing a spring. One such arrangement imagines forming for the poles to such an extent that every bar would have a precious stone like shape in cross-area, and with two sharp corners and two adjusted corners. The pole, as in this manner designed, consequently would have two sharp edges, similar to a sword, to guarantee that the bird will be parted into two pieces, or more relying upon where the effect happens.

When stowed the individuals or bars live in the front part of the motor nacelle. In one favored, endless supply of a control signal for incitation, gases created by little touchy charges move the ring forward to push the bars out of the nacelle. The ring is directed by straight heading and continues on a way corresponding to the motor shaft, that is, its longitudinal pivot. A proper ring is likewise mounted inside the nacelle near the delta lip and capacities to direct the poles or cylinders during the arrangement step. Openings in the ring serve to arrange the poles forward and towards the center. The poles, tubes or other molded individuals meet at or close to a point, shaping a cone.

At the foundation of the cone extra bar like individuals structure a cross section which secures the external annulus of the motor admission region. The individuals from this grid connect to the essential poles and to the nacelle lip. At the point when the gadget is to be withdrawn, the nacelle lip connect focuses discharge the optional poles. The ring to which the essential bars are connected is winched once again into its unique area in the nacelle through a cabled pulley framework with the bars being directed by the aide openings toward the front of the nacelle additionally return to their stowed position. The terms used to depict this substitute encapsulation have their customary significance as would be regularly perceived in this field and it is accepted that development of this epitome is inside the normal ability of this craftsmanship.

In one more favored epitome an inflatable pack is given and adjusted to be sent before the motor. The pack would be conveyed and expand quickly in much a similar way as vehicle airbags. Additionally, similarly as with the cone of poles encapsulation, the airbag epitome would be activated by the danger level evaluator sub-framework. As a result of an airbag's relatively more noteworthy effect on wind stream into the motor, it is alluring to limit the term of the sending, that is, to activate withdrawal at the earliest opportunity once the bird has been avoided. Two exemplifications or approaches are liked for organization, and they are dependent upon the accuracy with which the bird's area still up in the air. It is accepted that a bird's area, once followed by the sensor sub-framework, might be known distinctly to inside a range of a couple of meters.

To lessen this vulnerability a double sensor sub-framework would be utilized. In such a utilization an essential, nose mounted sensor would clear the airspace before the plane and would be utilized to decide onto which side of the fuselage the bird it prone to encroach. Following this assurance, a second, or last-second setting off sensor situated on the assumed effect side would affirm the bird's way to deal with that side and along these lines convey the airbag at the suitable motor.

Then again, if the direction of the approaching danger, attempted to be a bird, comparative with the airplane can be anticipated by the evaluator sub-framework with relative high assurance, the airbag's organization could in principle be focused on to slam into the bird on the way to the airbag's re-visitation of a stowed arrangement. In this setting the expression "airbag" is utilized comprehensively to allude to an adaptable as well as inflatable part that could play out the depicted capacity. Capturing the bird at the right moment would enormously diminish the measure of time that the sent gadget or part hinders the wind stream into the motor.

Two shapes are imagined for the airbag structures. One would have a minaret shape that might actually avoid the bird and put it on a way around the nacelle, as opposed to through the motor. Another shape is like that of the notable Hershey's kiss treats drops, with level end pointing forward from the motor and the sharp end appended to the center

These mechanical sub-frameworks are expected to work on the advanced high-sidestep proportion turbofan fly motor; in any case, they could be utilized on the turbojet motor too. Furthermore, the frameworks are not restricted to utilize exclusively on a fly motor; there are other airplane types in which they can be used for insurance against sway from birds. In particular, they can be utilized to secure the uncovered, weak parts of a helicopter rotor framework against dangers presented by generally enormous birds and different articles.

The as of now portrayed frameworks are expected to resolve serious issues that have not been tended to by known airplane motor assurance frameworks. In those frameworks, ordinarily some kind of screen gadget is forever fastened before the motor. This causes somewhere around two significant issues for any fly airplane motor: managing undesirable impacts made by the actual idea of the gadgets, such as ice arrangement, and disturbance of wind stream into the motor which thusly causes huge and inadmissible decrease in motor execution.

OUTCOME

According to claim1# the innovation is to a will be a robot bird flying controllability assessment framework and assessment technique and the assessment framework, a microchip B produces upgrade beat arrangements with progressively changed strength as per the preset standard, a chip An is responsible for GPS information examination and significant data record, the chip B helps out the chip An and the creation is the important test information assortment of a robot bird to be identified are finished in a coordinated way, the gathered information are handled through an information investigation arrangement of a PC prior to being changed over into a direction dependent on an electronic guide, if the robot bird has the flying controllability can decided based on the direction and the applicable data, and the assessment if the robot bird is qualified can be finished.

1. According to claim1,2# the innovation is to a far reaching assessment of the robot bird flying controllability can be understood, the earlier data needed by flight route can be obtained, the visual deficiency of ensuing flight route tests is decreased, a lot of invalid monotonous work is kept away from, the functioning effectiveness is improved enormously, and the robot bird reasonable advancement can be sped up.
2. According to claim1,2,3# the innovation is to the robot bird has a tremendous utilizing esteem. The innovation is a contrasted and the ground creature, the bird has a bigger dynamic space and quick speed, subsequently having a bigger application advantage clearly; the reasonable and potential application is extremely wide and an advanced fly motor delta security framework that ensures against huge birds and works independently, that is torpid in routine airplane tasks.
3. According to claim1,2,5# the innovation is to a consequently activates its defensive device(s) just at the quick mark of need, then, at that point returning it (them) to a non-meddling position, including a RADAR framework and a LIDAR framework which distinguish birds entering the planned flight way of the airplane, and quick PC carried out computational calculations that track and recognize those from the identified set that are (1) projected to enter a zone which would prompt ingestion by the motor, and (2) of a size huge enough that they would genuinely harm the motor whenever ingested.
4. According to claim1,2,4,6# the development is to a protective components housed in the motor nacelle cowling or focus center, or in the fuselage structure that are quickly incited in the blink of an eye prior to showing up at the focal point to safeguard, redirect, lessen the size of the moving toward bird to an OK mass, or obliterate it, and, that after the ingestion danger has passed, the gadgets are stowed/withdrawn or securely discarded.

REFERENCES

[1] Abhigna, B.S., Chakraborty, S., Kumar, V., Medha: A collaborative E-concept framework, Proceedings of the 2018 International Conference On Communication, Computing and Internet of Things, IC3IoT 2018, 2019, pp. 117–122, 8668127

[2] Abidin, S., Vadi, V.K., Rana, A., On Confidentiality, Integrity, Authenticity, and Freshness (CIAF) in WSN, Advances in Computer, Communication and Computational Sciences, 2020, pp 87-97

[3] Ahay, A., Choudhury, T., Kumar, V., Hierarchical Evolution of Digital Games, Proceedings of 2019 International Conference on Computational Intelligence and Knowledge Economy, ICCIKE 2019, 2019, pp. 381–386, 9004378

[4] Area: Communication, Frequency modulation, Networks, wireless communication

[5] Area: Communication, Frequency modulation, Networks, wireless communication

[6] Area: Communication, Frequency modulation, Networks, wireless communication

[7] Area: Communication, Frequency modulation, Networks, wireless communication

[8] Arora, M., Kumar, V., Dave, M., Hybrid cloud infrastructure for load balancing and scaling with vm path optimization using ant colony optimization, Journal of Advanced Research in Dynamical and Control Systems, 2020, 12(5 Special Issue), pp. 660–674

[9] Arora, M., Kumar, V., Dave, M., Task scheduling in cloud infrastructure using optimization technique genetic algorithm, Proceedings of the World Conference on Smart Trends in Systems, Security and Sustainability, WS4 2020, 2020, pp. 788–793, 9210303

[10] Bansal, R., Gupta, A., Singh, R., Nassa, V.K., Role and impact of digital technologies in E-learning amidst COVID-19 pandemic, Proceedings - 2021 4th International Conference on Computational Intelligence and Communication Technologies, CCICT 2021, 2021, pp. 194–202

[11] Bansal, R., Obaid, A.J., Gupta, A., Singh, R., Pramanik, S., Impact of Big Data on Digital Transformation in 5G Era, Journal of Physics: Conference Seriesthis link is disabled, 2021, 1963(1), 012170

[12] Chauhan, S.S., Khare, A.K., Kinematic analysis of the ABB IRB 1520 industrial robot using roboanalyzer software, Evergreenthis link is disabled, 2020, 7(4), pp. 510–518

[13] Chauhan, S.S., Khare, A.K., Method of Obtaining Planar State of Stress Using Mohr's Circle—Some Typical Cases, Lecture Notes in Mechanical Engineering, 2022, pp. 33–38

[14] Goel, S., Tripathi, A., Choudhury, T., Kumar, V. Parkinson's Disease Detection using Soft Computing Technique, Proceedings of the 2019 8th International Conference on System Modeling and Advancement in Research Trends, SMART 2019, 2020, pp. 3–6, 9117384

[15] Govindarajan, V., Sivakumar, R., Patil, P.P., Kannan, M., Ramesh, B., Effect of Tungsten Carbide Addition on the Microstructure and Mechanical Behavior of Titanium Matrix Developed by Powder Metallurgy Route, Advances in Materials Science and Engineeringthis link is disabled, 2022, 2022, 2266951

[16] Gupta, A., Singh, R., Nassa, V.K., Sharma, P., Koti, K., Investigating Application and Challenges of Big Data Analytics with Clustering, 2021 International Conference on Advancements in Electrical, Electronics, Communication, Computing and Automation, ICAECA 2021, 2021

[17] Jain, P., Vijay, S., & Gupta, S. C. (2009). ATM Network QoS Parameters and Performance Analysis using OPNET Simulator. International Journal of Recent Trends in Engineering, 2(4), 228.

[18] Jain, P., Vijay, S., & Gupta, S. C. Performance Analysis & QoS Guarantee in ATM Networks. Global Journal of computer Science and Technology, 131-136.

[19] Jain, S., Choudhury, T., Kumar, V., Kumar, P., Detecting malware and analysing using sandbox evasion, Proceedings of the 2018 International Conference On Communication, Computing and Internet of Things, IC3IoT 2018, 2019, pp. 111–116, 8668149

[20] Kala, S., Nandan, H., Sharma, P., Shadow and weak gravitational lensing of a rotating regular black hole in a non-minimally coupled Einstein-Yang-Mills theory in the presence of plasma, European Physical Journal Plusthis link is disabled, 2022, 137(4), 457

[21] Kandpal, H., Kumar, A., Malik, A., Morphometric analysis and prioritization of sub-watersheds in Bino watershed, Uttarakhand: A remote sensing and GIS perspective, Indian Journal of Ecology, 2018, 45(3), pp. 475–478

[22] Kannan, M., Singh, S., Prasad, R.R., Synthetic Methods to Obtain Calcia-Stabilized Zirconia Powders—A Review, Lecture Notes in Mechanical Engineeringthis link is disabled, 2021, pp. 405–416

[23] Kukreti, A., Painolli, A.K., Rana, N., Where do we stand: factors affecting sustainable development, Procedia Environmental Science, Engineering and Management, Volume 8, Issue 2, 2021, Pages 583-589

[24] Kulkarni, S & Chima, P 2021 'Challenges faced by UK university students due to thcoronavirus crisis in the Higher Education' Preprints.org. https://doi.org/10.20944/preprints202102.0192.v1

[25] Kulkarni, S & Sodre, JR 2017, An Assessment of Lightweight Vehicles to Reduce Greenhouse Gas Emissions with Focus on Magnesium. in Thoughts and Refections on the use of Magnesium. Birmingham City University, pp. 84-91.

[26] Kulkarni, S, Chapman, C & Shah, H 2016 'Computational Fluid Dynamics (CFD) Mesh Independency Study of A Straight Blade Horizontal Axis Tidal Turbine' Preprints.org. https://doi.org/10.20944/preprints201608.0008.v1

[27] Kulkarni, S, Chapman, C, Shah, H & Edwards, D 2018, 'A computational design method for bio-mimicked horizontal axis tidal turbines', International Journal of Building Pathology and Adaptation, vol. 36, no. 2, pp. 188-209. https://doi.org/10.1108/IJBPA-06-2017-0029

[28] Kulkarni, S, Chapman, C, Shah, H, Parn, E & Edwards, DJ 2018, 'Designing an efficient tidal turbine blade through bio-mimicry: A systematic review', Journal of Engineering, Design and Technology, vol. 16, no. 1, pp. 101-124. https://doi.org/10.1108/JEDT-08-2017-0077

[29] Kulkarni, S, Chapman, C, Shah, H, Parn, E & Edwards, DJ 2018, 'Design Study of Horizontal Axis Tidal Turbine Blade', Mindanao Journal of Science and Technology. https://www.open-access.bcu.ac.uk/10089/

[30] Kulkarni, S, Edwards, DJ, Chapman, C, Hosseini, MR & Owusu-Manu, D-G 2019, 'A preliminary mechanical design evaluation of the Wikispeed

car: for light-weighting implications', Journal of Engineering, Design and Technology, vol. 17, no. 1, pp. 230-249. https://doi.org/10.1108/JEDT-09-2018-0154

[31] Kulkarni, S, Edwards, DJ, Parn, E, Chapman, C, Aigbavboa, CO & Cornish, R 2018, 'Evaluation Of Vehicle Lightweighting To Reduce Greenhouse Gas Emissions With Focus On Magnesium Substitution', Journal of Engineering, Design and Technology, vol. 16, no. 6, pp. 869-888. https://doi.org/10.1108/JEDT-03-2018-0042

[32] Kulkarni, SS, Wang, L, Golsby, N & Lander, M 2021, 'Fluid-structure interaction based optimisation in tidal turbines: A perspective review', Journal of Ocean Engineering and Science. https://doi.org/10.1016/j.joes.2021.09.017

[33] Malik, A., Kumar, A., Kandpal, H., Morphometric analysis and prioritization of sub-watersheds in a hilly watershed using weighted sum approach, Arabian Journal of Geosciencesthis link is disabled, 2019, 12(4), 118

[34] Mohanakurup, V.K., Baghela, V.S., Kumar, S., Srivastava, P.K., Doohan, N.V., Soni, M., Awa, H., 5G Cognitive Radio Networks Using Reliable Hybrid Deep Learning Based on Spectrum Sensing, Green Internet of Things in 5G and Beyond, Hindawi, Volume 2022

[35] Mohta, M., Soni, N., Choudhury, T., Kumar, V., Market Analysis on Li-Fi Technology, Proceedings of the 2019 8th International Conference on System Modeling and Advancement in Research Trends, SMART 2019, 2020, pp. 139–143, 9117427

[36] Nagwan, H., Rana, N., Kumar, A., Tewari, P.C., Performance and emission characteristics of diesel engine using Karanja oil methyl ester and Eucalyptus oil fuel blends, Procedia Environmental Science, Engineering and Management, Volume 6, Issue 4, 2019, Pages 553-563

[37] Painoli, A.K., Bansal, R., Singh, R., Kukreti, A., Impact of digital marketing on the buying behavior of youth with special reference to Uttarakhand state, Big Data Analytics for Improved Accuracy, Efficiency, and Decision Making in Digital Marketing, 2021, pp. 162–182

[38] Parag, J. (2011). Hybrid mixed handover for call blocking probability analysis in wireless ATM network. Journal of Engineering and Computer Innovations, 2(4), 68-77.

[39] Sharma, C., Singh, M.P., Chaudhary, S., Cost Analysis of Solar-Wind Based Hybrid Renewable Energy System, Lecture Notes in Electrical Engineeringthis link is disabled, 2021, 700, pp. 2733–2746

[40] Sharma, S., Chandra, U., & Jain, P. (2017). A literature survey on automation of test data generation for branch coverage testing using genetic algorithm. International Journal of Computational Intelligence Research, 13(6), 1521-1531.

[41] Varshney, R., Jain, P., & Vijay, S. (2018, September). Massive MIMO Systems In Wireless Communication. In 2018 2nd International Conference on Micro-Electronics and Telecommunication Engineering (ICMETE) (pp. 39-44). IEEE.

CHAPTER - 15

Data Analysis and Visualization Using AI

Dr. Anil W. Kale

Mahatma Gandhi Mission's College of Engineering and Technology, India. Email: anil5474@gmail.com

Dr. Nandkishor P Karlekar

MGMCET Kamothe Navimumbai, India. Email: npkarlekar.mgmcet@gmail.com

Dr. Vaibhav Narawade

Ramrao Adik Institute of Technology, India. Email:vnarawade@gmail.com

Dr. Sanjay B. Waykar

Mahatma Gandhi Mission's College of Engineering and Technology, India. Email:sbwaykar@gmail.com

ABSTRACT

With regards to information perception and examination, this report traces a portion of the difficulties and arising applications that emerge in the Big Data time. In especially, fourteen recognized researchers from the scholarly world and industry, and different related networks, i.e., Information Visualization, Human-Computer Interaction, Machine Learning, Data the board and Mining, and Computer Graphics have been welcome to offer their viewpoints. Information perception is a very new and promising field in software engineering. It utilizes PC realistic impacts to uncover the examples, patterns, connections out of datasets. In this paper, we initially get to know information perception and its connected ideas, then, at that point we will glance through some broad calculations to do the information representation. To get further with regards to it, we will have some conversation about multidimensional information perception. With the mix of some known strategies, we present another calculation to do 4 dimensional information perception. We likewise present a program project plan about it (discretionary), and a few issues and clarifications around it.

INTRODUCTION

Perception is the graphical show of data, fully intent on giving the watcher a subjective comprehension of the data substance. It is likewise the method involved with changing items, ideas, and numbers into a structure that is apparent to the natural eyes. At the point when we say "data", we might allude to information, cycles, relations, or ideas. Here, we confine it to information.

Information perception and investigation are these days one of the foundations of Data Science, turning the bounty of Big Data being delivered through present day frameworks into significant information. Without a doubt, the Big Data period has understood the accessibility of voluminous datasets that are dynamic, uproarious and heterogeneous in nature. Changing an information inquisitive client into somebody who can get to and investigate that information is much more troublesome now for an incredible number of clients with next to zero help and skill on the information handling part.

Consequently, the space of information perception and investigation has acquired extraordinary consideration as of late, calling for joint activity from various examination regions from the Information Visualization, Human-Computer Interaction, Machine Learning, Data the executives and Mining, and Computer Graphics. A few conventional issues from those networks, like proficient information stockpiling, questioning and ordering for empowering visual investigation, ways for visual show of enormous information, effective communication and personalization methods that can fit to various client needs, are returned to in view of Big Data.

This is to empower present day perception frameworks that offer versatile procedures to proficiently deal with billion articles datasets, while restricting the visual reaction to a couple of milliseconds. In this report, with regards to the third International Workshop on Big Data Visual Exploration and Analytics, the getting sorted out board welcomed fourteen recognized researchers, from various networks to give their bits of knowledge in regards to the difficulties and the applications they observe to be more fascinating in coming years, identified with Big information perception and analytics2.

Especially, every researcher sums up his musings with respect to the accompanying two angles: – the top future examination challenges in Big Data perception and investigation – the top arising applications with regards to Big Data representation and investigation We present their reactions in the accompanying segments, while the difficulties are summed up as follows.

Initial, a test identified with the Machine Learning (ML) field is empowering intuitive AI. The test is to assemble intelligent apparatuses and perceptions for AI that can give client proposals and backing client driven ML applications. Likewise, it includes intuitive strategies for translation, troubleshooting, and examinations of ML models. One more test is identified with the versatility and proficiency of information representation and communication activities.

Toward that path, the difficulties include how to construct devices that can perform intelligent activities and complex examination over gigantic arrangements of information. In that regard, there is the requirement for novel methodologies (e.g., reformist information preparing) that can effectively deal with enormous streaming, examined, questionable, high-dimensional, loud information.

Information representation is tied in with getting proportions and connections among numbers. Not tied in with understanding individual numbers, but rather about understanding the examples, patterns, and connections that exist in gatherings of numbers. From the purpose in client understanding, it might include discovery, estimation, and correlation, and is improved by means of intelligent strategies and giving the data from various perspectives and with different methods.

For what reason do we do information perception? To see and comprehend pictures is one of the normal senses of human, and to comprehend mathematical information is a year's preparation expertise from schools, and all things considered, a many individuals are as yet bad with mathematical information. From an all-around drawn picture, one is a lot simpler to discover the patterns and relations. Since visual show of data exploits the immense, and frequently underutilized, limit of the natural eye to distinguish data from pictures and outlines. Information representation moves the heap from mathematical thinking to visual thinking.

Getting data from pictures is undeniably additional efficient than glancing through message and numbers – that is the reason numerous leaders would prefer to have data introduced to them in graphical structure, instead of a composed or printed structure. Something else we should specify is that: information perception isn't logical representation. Logical perception utilizes liveliness, recreation, and refined PC illustrations to make visual models of designs and handled that can't in any case be seen, or seen in adequate detail.

While information representation is a way that present and show data in a manner that empowers suitable translation, choice, and affiliation. It uses human abilities for design acknowledgment and pattern investigation, and takes advantage of the capacity of individuals to remove a lot of data in a brief timeframe from visuals introduced in a normalized design.

SUMMARY

Gennady and Natalia Adriano Visual Analytics for Data Science: A Critical View. Visual portrayals are frequently utilized in information examination. In conventional information mining draws near, representations show up at the finish of scientific work processes, focusing on understanding of recognized examples and their correspondence to different beneficiaries, e.g., different experts, chiefs or overall population. As per the visual examination theory while human endeavors should be diminished however much as could be expected by computational handling, perception should be utilized all through the whole insightful work process at whatever point an examiner should make educated choices concerning further strides.

The job of representation is to pass on the fundamental data to the human in a structure empowering successful insight and comprehension. Henceforth, the errand of visual investigation is to foster logical work processes in which human perception is adequately upheld by representations and computational preparing. Visual investigation ought to likewise be engaged with improvement of new calculations and programming instruments for computerized examination and displaying (e.g., AI techniques) for checking whether and how well the strategies are getting along what they are planned to do. Future Challenges. Our significant examination subject is outwardly determined investigation of spatio-worldly information.

An agent model is development information comprising of arrangements of time-referred to positions. An assortment of strategies and instruments exist as of now (see hypothetical establishments in the book; new techniques are growing effectively (this is affirmed by an enormous number of acknowledged papers on this point at IEEE VIS and other profoundly specific gatherings) and applied effectively in such spaces as transportation and game investigation.

While the accompanying considerations have emerged from our encounters with spatio-worldly information, they are material to different kinds of intricate information. Arising Applications. A few key late advancements set out incredible open doors for enabling information science by visual examination. The first is the appearance and wide spread of information science-situated dialects like Python and R. These dialects empower bit by bit information examination and backing joining of perceptions in logical work processes. Logical scratch pad dependent on these dialects help to report examination, empower reproducibility and help to share results.

These advancements are so natural to utilize that these days everybody can turn into an examiner. There exist course readings clarifying the nuts and bolts on basic models, and many bits of code are accessible online for use and variation. This wonder has its rear: independent examiners frequently need essential information on the general examination cycle, and miss comprehension of why, when, and how perceptions should be utilized in investigation.

The Internet is stuffed by perceptions, frequently looking exceptionally great and extravagant, that convey fake examples insufficiently. Shockingly, the greater part of the accessible code models and most of reading material don't go past applying fundamental illustrations to straightforward information and don't exhibit the logical worth of the designs. Accordingly, we see the significant test in teaching information researchers on the most proficient method to utilize perceptions accurately and adequately inside non-inconsequential logical work processes, understanding and considering the information types, properties of the information, and their intricacy.

Future Challenges and Emerging Applications. Frameworks for Machine Learning (ML) and ML for Systems. While maybe overhyped, the immense measure of consideration coordinated towards Machine Learning is obvious all through exploration and industry. Papers are seeming covering various angles, from principal hypothetical advances like causal thinking and general insight to uses of AI in frameworks and information work.

This blast in AI is empowered essentially by the tremendous measures of information accessible and frameworks that permit the preparation for models utilizing those huge informational indexes. This thusly empowers astute uses of these strategies far in excess of the fundamental wildernesses of ML (arrangement, grouping, and relapse). Given this blast, we need much better procedures for working with the information and the models for ML.

This incorporates investigating models, getting where models work and don't work, contrasting models and one another, and giving reasonable clarifications for model conduct. Since perception is generally about assisting people with interpreting and cooperate with information, intuitive devices and representations for AI is a one of the top difficulties for perception in the coming decade.

Simultaneously, utilizing the yield of models to assist work with bettering representations (regardless of whether it's for suggesting a solitary or arrangement of perceptions) or to collaborate at a more significant level with information by aiding discover ideal methods of utilizing human instinct and information while taking advantage of all the more remarkable calculation is a critical part of new applications.

DIAGRAM

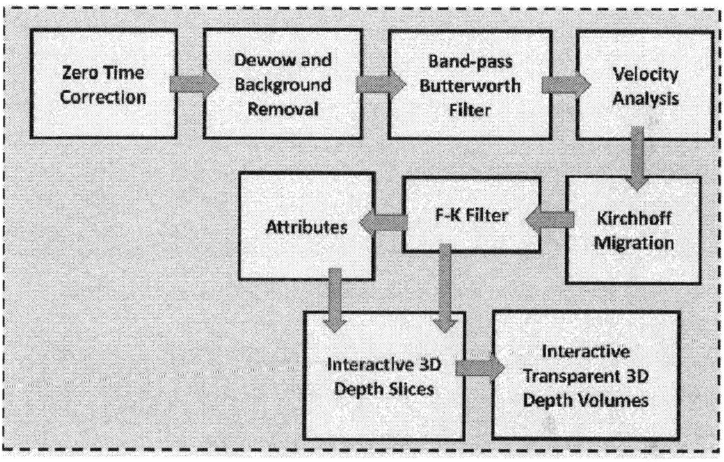

Fig. 1: Data Analysis and Visualization, Flow Chart

Our second significant objective was to investigate the impact of the shot ATMs on the dirt qualities to comprehend soil pollu-tion. Landmines might be the most boundless, deadly, and dependable type of contamination on the planet. This article additionally inspects the dirt contaminations by confocal Raman spectrometer (CRS) and enraptured energy dispersive X-beam fluorescence (PEDXRF) techniques in the wake of shooting these caught weapons in the land. CRS and PEDXRF analyzing procedures are critical in this specific circumstance, and are notable techniques for the investigation of minerals and compound organizations of the dirt's including poisonous components, colors, semiconductor materials, natural constructions, and polymers. This can be accomplished to acquire the large scale and minor component structures of the dirts and rocks. The consequences of the CRS and PEDXRF examinations could expose where the ATMs ought to have been exploded, regardless of whether they should be conveyed to the unplanted and automated secure region or still up in the air area nearby.

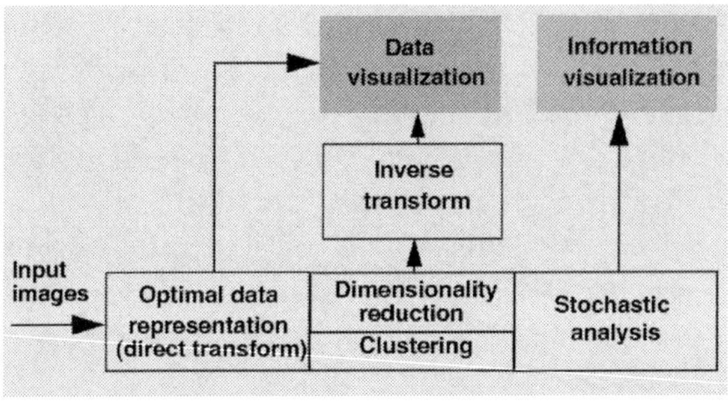

Fig. 2: Data Analysis and Visualization, Block Diagram

Square chart of an information and data mining device. The apparatus permits the intuitive investigation of the arrangement of pictures acquired as a question reaction. The two information and data representation techniques will be utilized after intuitive pursuit of pictures significant in a specific application.

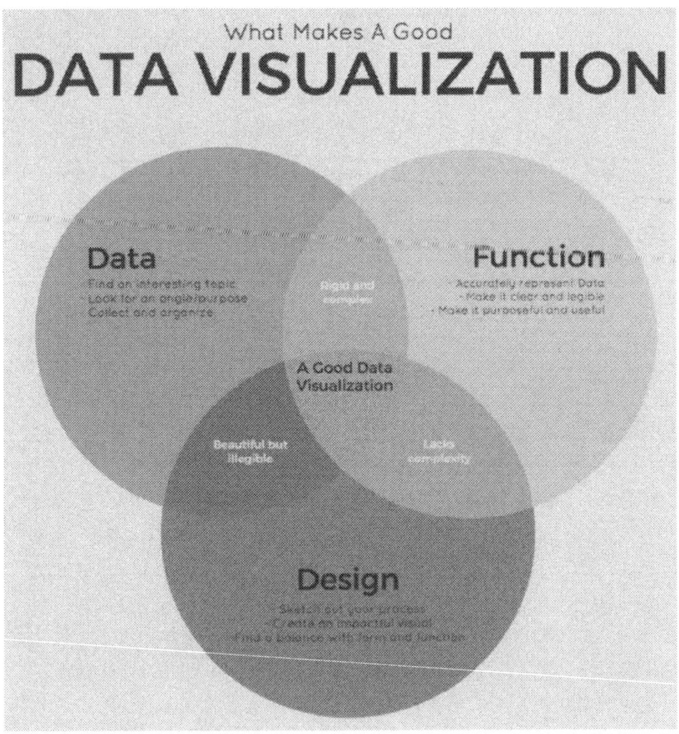

Fig. 3: Data Analysis and Visualization

DESCRIPTION

Future Challenges. To be viable, perception and visual examination ought to be intuitive, implying that registering visual portrayals ought to occur in no time flat, communicating on them ought to be responsive, and investigation ought to likewise be done as per the satisfactory furthest reaches of human idleness as depicted in the writing.

Building frameworks that stay intelligent at scale and utilizing complex examination is quite difficult for the representation field, which might become superfluous in the event that it doesn't address the adaptability challenge appropriately. Arising Applications. To address this versatility challenge, my new focal point of exploration is "Reformist Data Analysis and Visualization", another worldview of calculation that, rather than performing calculations in a single step that can require some investment to finish, parts them in a progression of short pieces of inexact calculations that improve with time.

In this way, rather than sitting tight for an unbounded measure of time the consequences of calculations for representation and investigation, experts can see the outcomes unfurling dynamically. They can, in this way, keep up with their consideration and begin settling on certain choices sooner than if they needed to trust that the entire calculations will wrap up. In the meantime, while the outcomes are being processed, investigators can likewise cooperate with the continuous calculation, changing calculation boundaries and in some cases controlling the calculations.

This new worldview is simply beginning to arise and will require more opportunity to become standard, as clarified in the report we distributed after a Dagstuhl workshop led in 2018. Nonetheless, I am sure that Progressive Data Analysis will permit perception and examination to turn out to be more adaptable while staying intelligent to work with the investigation of the abundance of information that the world is gathering, combined with new incredible techniques to dissect it created in AI specifically.

This new worldview isn't just significant for representation and visual examination yet additionally requires solid joint efforts with scientists in Databases and Machine Learning who perceive that reformist information investigation will prompt more adaptable exploratory frameworks. Arising Applications. We are figuring out how to request new things from our information. It's undeniably functional to intelligently investigate Big Data, posing novel inquiries to find unforeseen wonders.

The lines between various types of examination - from social inquiries, to unstructured information, to rich media are obscuring. I'm anticipating our working on all means of the cycle: to figuring out how to best communicate questions; how to get intelligent speed reactions to those inquiries; and to repeat on those experiences to ask the following round of inquiries. These means are interconnected and related.

Space overall, that is gigantic and threatening. Luckily, I can focus on my work on Honeycomb5 toward their genuine issues, thus can exploit the limitations of their specific setting. Honeycomb is an APM (Application Performance Monitoring) instrument for troubleshooting appropriated frameworks. Our clients are DevOps, who are liable for sending new code - and recuperating when it falls flat.

I accept that zeroing in determinedly on explicit use cases will simplify in any case wide inquiries. In the event that we can truly get what the client needs to tackle their concern, we can realize what kinds of information they will ingest, what questions they should ask, what execution qualities they expect, and what level of accuracy they need.

Imagining information is one of the most outstanding approaches to discover examples and data in Big Data. The motivation behind why information perception is fascinating for the information the board local area is that this is an intrinsically information escalated issue. Moreover, information researchers might act subjective inquiries like they make new representations or associate with existing ones. This implies that such frameworks get:

1. diverse queries,
2. sequences of queries where each query may depend on the previous one,
3. queries that may be OK to abort,
4. workloads which need rapid response times to remain interactive even if correctness is not immediately at 100%.

Because of this crisscross with normal data set applications, there are a few long haul and invigorating difficulties that address awesome freedoms for information the board scientists given the rich history of the field in information escalated calculations and frameworks. I feature two of those chances as they emerge from ongoing work nearby.

To start with, what is what might be compared to the social variable based math for visual investigation? It may appear to be overwhelming to gather the immense space of potential activities an information researcher might perform into a little arrangement of tasks, however this is actually what the first social variable based math accomplished. And afterward, more perplexing tasks can be combined from crude polynomial math activities.

On top of that polynomial math, we would then be able to construct frameworks that depend on a typical interface and a little arrangement of administrators, permitting the local area to aggregately tackle this issue by considering elective plans and executions that regard similar model and API as it occurred with social administrators.

This reflection is one of the mysteries both for the reception of the social model across different applications and for the capacity to generally effectively try different things with elective executions. - Second, what is what could be compared to the b-tree, and the roworiented and section situated capacity plans? While these are in no way, shape or form the main ordering and capacity choices, knowing the outrageous plans or probably the most adaptable plans, and afterward vigorously zeroing in on them, permitted social data sets to develop as far as both speed and power. Information representation and visual examination is (regularly) an information serious rather than a figure concentrated issue. This implies that the manner in which we store and move information is the bottleneck.

Future Challenges. - Interactive information investigation for huge information and more intricate activities. Apparatuses like Tableau, Power Bi and co applause themselves as intelligent information investigation devices. However, for bigger datasets they depend on pre-registered information 3D squares, appeared sees, and comparative methods to remain intuitive. Shockingly, these strategies seriously limit what the client can do.

For instance, to make an information block one has to know forthright, what kind of inquiries the information solid shape should reply. This basically forestalls intuitive reactions for totally new inquiries. A key examination challenge is, the way we can assemble frameworks which ensure intuitive reaction times paying little heed to the inquiry and information size. As a feature of North star, we began to investigate how we can use reformist calculation and testing to accomplish this. - Sustainable experiences and understanding suggestion. We need to make discovering experiences simpler.

Hence, we need to foster devices which help non-Data Scientists to find experiences and ceaselessly screen the information for changes. For instance, a framework ought to consequently suggest intriguing experiences and representation about things that may intrigue the client. See-DB or Viz-ML are frameworks, what began to investigate that. Simultaneously, those experiences ought to likewise be practical. For instance, current knowledge proposal frameworks generally disregard the danger of discovering false experiences by testing such a large number of theories.

Novel interfaces. We should make information investigation more available to a more extensive scope of clients. This needs to essentially reevaluate the UI. We put everything on the table HCI has to bring to the table; from novel perceptions, collaboration designs, contact screens, up to normal language interfaces. Curiously, changing the UI frequently additionally has serious ramifications on how the backend must be created.

We (the SIGMOD people group) will in general initially foster the framework and afterward add the UI as an idea in retrospect, regularly prompting inconvenient old-style cooperation. I figure it ought to be the opposite way around. Plan the client connections first and afterward sort out the framework, which can really uphold them.

REFERENCES

[1] Abhigna, B.S., Chakraborty, S., Kumar, V., Medha: A collaborative E-concept framework, Proceedings of the 2018 International Conference On Communication, Computing and Internet of Things, IC3IoT 2018, 2019, pp. 117–122, 8668127

[2] Abidin, S., Vadi, V.K., Rana, A., On Confidentiality, Integrity, Authenticity, and Freshness (CIAF) in WSN, Advances in Computer, Communication and Computational Sciences, 2020, pp 87-97

[3] ahay, A., Choudhury, T., Kumar, V., Hierarchical Evolution of Digital Games, Proceedings of 2019 International Conference on Computational Intelligence and Knowledge Economy, ICCIKE 2019, 2019, pp. 381–386, 9004378

[4] Area: Communication, Frequency modulation, Networks, wireless communication

[5] Area: Communication, Frequency modulation, Networks, wireless communication

[6] Area: Communication, Frequency modulation, Networks, wireless communication

[7] Area: Communication, Frequency modulation, Networks, wireless communication

[8] Arora, M., Kumar, V., Dave, M., Hybrid cloud infrastructure for load balancing and scaling with vm path optimization using ant colony optimization, Journal of Advanced Research in Dynamical and Control Systems, 2020, 12(5 Special Issue), pp. 660–674

[9] Arora, M., Kumar, V., Dave, M., Task scheduling in cloud infrastructure using optimization technique genetic algorithm, Proceedings of the World

Conference on Smart Trends in Systems, Security and Sustainability, WS4 2020, 2020, pp. 788–793, 9210303

[10] Bansal, R., Gupta, A., Singh, R., Nassa, V.K., Role and impact of digital technologies in E-learning amidst COVID-19 pandemic, Proceedings - 2021 4th International Conference on Computational Intelligence and Communication Technologies, CCICT 2021, 2021, pp. 194–202

[11] Bansal, R., Obaid, A.J., Gupta, A., Singh, R., Pramanik, S., Impact of Big Data on Digital Transformation in 5G Era, Journal of Physics: Conference Seriesthis link is disabled, 2021, 1963(1), 012170

[12] Chauhan, S.S., Khare, A.K., Kinematic analysis of the ABB IRB 1520 industrial robot using roboanalyzer software, Evergreenthis link is disabled, 2020, 7(4), pp. 510–518

[13] Chauhan, S.S., Khare, A.K., Method of Obtaining Planar State of Stress Using Mohr's Circle—Some Typical Cases, Lecture Notes in Mechanical Engineering, 2022, pp. 33–38

[14] Goel, S., Tripathi, A., Choudhury, T., Kumar, V. Parkinson's Disease Detection using Soft Computing Technique, Proceedings of the 2019 8th International Conference on System Modeling and Advancement in Research Trends, SMART 2019, 2020, pp. 3–6, 9117384

[15] Govindarajan, V., Sivakumar, R., Patil, P.P., Kannan, M., Ramesh, B., Effect of Tungsten Carbide Addition on the Microstructure and Mechanical Behavior of Titanium Matrix Developed by Powder Metallurgy Route, Advances in Materials Science and Engineeringthis link is disabled, 2022, 2022, 2266951

[16] Gupta, A., Singh, R., Nassa, V.K., Sharma, P., Koti, K., Investigating Application and Challenges of Big Data Analytics with Clustering, 2021 International Conference on Advancements in Electrical, Electronics, Communication, Computing and Automation, ICAECA 2021, 2021

[17] https://www.tableau.com/learn/articles/data-visualization#:~:text=Data%20visualization%20is%20the%20graphical,outliers%2C%20and%20patterns%20in%20data.

[18] https://www.udacity.com/course/data-analysis-and-visualization--ud404

[19] https://mealdprostarter.org/n-data-analysis-visualization-and-interpretation/

[20] https://www.coursera.org/learn/data-analyze-visualize

[21] https://www.coursera.org/specializations/data-analysis-visualization-foundations

[22] https://www.linkedin.com/learning/data-visualization-for-data-analysis-and-analytics

[23] Jain, P., Vijay, S., & Gupta, S. C. (2009). ATM Network QoS Parameters and Performance Analysis using OPNET Simulator. International Journal of Recent Trends in Engineering, 2(4), 228.

[24] Jain, P., Vijay, S., & Gupta, S. C. Performance Analysis & QoS Guarantee in ATM Networks. Global Journal of computer Science and Technology, 131-136.

[25] Jain, S., Choudhury, T., Kumar, V., Kumar, P., Detecting malware and analysing using sandbox evasion, Proceedings of the 2018 International Conference On Communication, Computing and Internet of Things, IC3IoT 2018, 2019, pp. 111–116, 8668149

[26] Kala, S., Nandan, H., Sharma, P., Shadow and weak gravitational lensing of a rotating regular black hole in a non-minimally coupled Einstein-Yang-Mills theory in the presence of plasma, European Physical Journal Plusthis link is disabled, 2022, 137(4), 457

[27] Kandpal, H., Kumar, A., Malik, A., Morphometric analysis and prioritization of sub-watersheds in Bino watershed, Uttarakhand: A remote sensing and GIS perspective, Indian Journal of Ecology, 2018, 45(3), pp. 475–478

[28] Kannan, M., Singh, S., Prasad, R.R., Synthetic Methods to Obtain Calcia-Stabilized Zirconia Powders—A Review, Lecture Notes in Mechanical Engineeringthis link is disabled, 2021, pp. 405–416

[29] Kukreti, A., Painolli, A.K., Rana, N., Where do we stand: factors affecting sustainable development, Procedia Environmental Science, Engineering and Management, Volume 8, Issue 2, 2021, Pages 583-589

[30] Kulkarni, S & Chima, P 2021 'Challenges faced by UK university students due to thcoronavirus crisis in the Higher Education' Preprints.org. https://doi.org/10.20944/preprints202102.0192.v1

[31] Kulkarni, S & Sodre, JR 2017, An Assessment of Lightweight Vehicles to Reduce Greenhouse Gas Emissions with Focus on Magnesium. in Thoughts and Refections on the use of Magnesium. Birmingham City University, pp. 84-91.

[32] Kulkarni, S, Chapman, C & Shah, H 2016 'Computational Fluid Dynamics (CFD) Mesh Independency Study of A Straight Blade Horizontal Axis Tidal Turbine' Preprints.org. https://doi.org/10.20944/preprints201608.0008.v1

[33] Kulkarni, S, Chapman, C, Shah, H & Edwards, D 2018, 'A computational design method for bio-mimicked horizontal axis tidal turbines', International Journal of Building Pathology and Adaptation, vol. 36, no. 2, pp. 188-209. https://doi.org/10.1108/IJBPA-06-2017-0029

[34] Kulkarni, S, Chapman, C, Shah, H, Parn, E & Edwards, DJ 2018, 'Designing an efficient tidal turbine blade through bio-mimicry: A systematic review', Journal of Engineering, Design and Technology, vol. 16, no. 1, pp. 101-124. https://doi.org/10.1108/JEDT-08-2017-0077

[35] Kulkarni, S, Chapman, C, Shah, H, Parn, E & Edwards, DJ 2018, 'Design Study of Horizontal Axis Tidal Turbine Blade', Mindanao Journal of Science and Technology. https://www.open-access.bcu.ac.uk/10089/

[36] Kulkarni, S, Edwards, DJ, Chapman, C, Hosseini, MR & Owusu-Manu, D-G 2019, 'A preliminary mechanical design evaluation of the Wikispeed car: for light-weighting implications', Journal of Engineering, Design and Technology, vol. 17, no. 1, pp. 230-249. https://doi.org/10.1108/JEDT-09-2018-0154

[37] Kulkarni, S, Edwards, DJ, Parn, E, Chapman, C, Aigbavboa, CO & Cornish, R 2018, 'Evaluation Of Vehicle Lightweighting To Reduce Greenhouse Gas

Emissions With Focus On Magnesium Substitution', Journal of Engineering, Design and Technology, vol. 16, no. 6, pp. 869-888. https://doi.org/10.1108/JEDT-03-2018-0042

[38] Kulkarni, SS, Wang, L, Golsby, N & Lander, M 2021, 'Fluid-structure interaction based optimisation in tidal turbines: A perspective review', Journal of Ocean Engineering and Science. https://doi.org/10.1016/j.joes.2021.09.017

[39] Malik, A., Kumar, A., Kandpal, H., Morphometric analysis and prioritization of sub-watersheds in a hilly watershed using weighted sum approach, Arabian Journal of Geosciencesthis link is disabled, 2019, 12(4), 118

[40] Mohanakurup, V.K., Baghela, V.S., Kumar, S., Srivastava, P.K., Doohan, N.V., Soni, M., Awa, H., 5G Cognitive Radio Networks Using Reliable Hybrid Deep Learning Based on Spectrum Sensing, Green Internet of Things in 5G and Beyond, Hindawi, Volume 2022

[41] Mohta, M., Soni, N., Choudhury, T., Kumar, V., Market Analysis on Li-Fi Technology, Proceedings of the 2019 8th International Conference on System Modeling and Advancement in Research Trends, SMART 2019, 2020, pp. 139–143, 9117427

[42] Nagwan, H., Rana, N., Kumar, A., Tewari, P.C., Performance and emission characteristics of diesel engine using Karanja oil methyl ester and Eucalyptus oil fuel blends, Procedia Environmental Science, Engineering and Management, Volume 6, Issue 4, 2019, Pages 553-563

[43] Painoli, A.K., Bansal, R., Singh, R., Kukreti, A., Impact of digital marketing on the buying behavior of youth with special reference to Uttarakhand state, Big Data Analytics for Improved Accuracy, Efficiency, and Decision Making in Digital Marketing, 2021, pp. 162–182

[44] Parag, J. (2011). Hybrid mixed handover for call blocking probability analysis in wireless ATM network. Journal of Engineering and Computer Innovations, 2(4), 68-77.

[45] Sharma, C., Singh, M.P., Chaudhary, S., Cost Analysis of Solar-Wind Based Hybrid Renewable Energy System, Lecture Notes in Electrical Engineeringthis link is disabled, 2021, 700, pp. 2733–2746

[46] Sharma, S., Chandra, U., & Jain, P. (2017). A literature survey on automation of test data generation for branch coverage testing using genetic algorithm. International Journal of Computational Intelligence Research, 13(6), 1521-1531.

[47] Varshney, R., Jain, P., & Vijay, S. (2018, September). Massive MIMO Systems In Wireless Communication. In 2018 2nd International Conference on Micro-Electronics and Telecommunication Engineering (ICMETE) (pp. 39-44). IEEE.

Made in the USA
Las Vegas, NV
19 September 2026

4b17f6fa-e072-4d70-adb0-9ba25d465c0fR01